Practical Veganism

Reaching for What Should Not Exceed Our Grasp – A Doable Alternative to "Impractical" Veganism

By Richard Fiordo* (PhD, CPT) and Dr. Noah Chelliah** (MD, FACC, FACP, FCCP)

* To be clear, Richard Fiordo (a coauthor of this book, Professor Emeritus from the University of North Dakota, Honorary Chair in Health Communication at Ambrosiana University, and a CPT or certified personal trainer) will be referred to as Fiordo. Fiordo acknowledges his lifelong debt to Robert Gajda, fitness expert and physical therapist, for advancing the PHA method of resistance training to him and the world.

** Dr. Noah Chelliah (a coauthor of this book, a cardiologist, President of the Heart Institute of North Dakota in Grand Forks, North Dakota, and Professor of Medicine at the Medical School at the University of North Dakota) will be referred to as either Dr. Chelliah or simply Chelliah. Dr. Chelliah wishes to acknowledge the wealth of information he obtained from the materials published by Dr. Michael Greger and would like to thank Dr. Greger for mentoring him.

Printed in the United States of America
ISBN 978-1-7377968-4-8

Dedication

To all human beings who want to live longer and healthier lives.

Contents

Preface

In this book on practical veganism (PV), we wish to counteract kneejerk reactions opposed to veganism. We distinguish practical veganism from impractical veganism. While we do not wish to disparage veganism, we accept that, optimistically, around five percent of the world population follows a vegan diet. There is no effort on our part to desacralize veganism and camouflage what is sacred about veganism with anything profane (Eliade, 1978). We know that to set veganism up as an ideal that all should emulate is to set up a standard that is impractical and unworkable for the supermajority of the human population. There is a witty saying about quitting the powerful habit of smoking that applies to veganism. "It is easy to quit smoking," says the comic, "I've done it a thousand times." Similarly, another comic might say about the powerful habit of eating meat: "It is easy to quit eating meat, I've done it a thousand times." One serious question emerges from efforts to live by a total vegan diet. How long can most people last on a diet that is completely vegan? The question has to do with chronemics or the study of time (DeVito, 2021) in relation to human activity. Most people can last several days with properly prepared vegan meals. However, to make veganism a lifestyle is a culinary time-trial that most people cannot sustain. Moose, deer, elephants, gorillas, and other creatures great and small sustain vegan lifestyles. Some people can too; however, proportionately, radical veganism is sustained by a mere human minority.

On medical and scientific grounds, the authors of this book aim to establish an *attainable and sustainable* practical veganism as a workable and tasty diet that has been practiced for centuries without being called by this name. Dr. Chelliah generated the useful and guiding term of "practical veganism." Our aim here is to share the practical vegan lifestyle with you so that you can see how healthy, manageable, accomplishable, and flavorful it can be. Now, more than ever, markets abound in vegan and vegan-like eating options. Since few, statistically speaking, will choose to be or can be thorough vegans, all of us can be practical vegans with advantages in lifespan and healthspan. As a cuisine chef from Tokyo, Japan noted to Fiordo: "Practical veganism sounds reasonable and possible to achieve for most people. Total veganism seems too impractical for most people to practice or embrace."

The American philosopher Charles W. Morris (1964; Fiordo, 1977) argued that scientific evidence should be the base for decision making and action. However, many decisions must be made and actions must be taken with limited scientific evidence to ground the decision and action. Morris maintained that when scientific evidence is limited or absent, a decision and action may still be demanded. When such circumstances prevail, we must do our best to draw educated and intelligent conclusions based on trustworthy experience that matters and whatever scientific data may be available. Although varying degrees of uncertainty may be present in the scientific findings, interpolation and extrapolation of scientific variables may be used to estimate what is not known.

When the American rhetorical scholar Richard M. Weaver (1948) wrote *Ideas Have Consequences* (1948), he had in mind ideas that covered many realms. What we are borrowing from Weaver's general idea is the predictable fact that dietary ideas have medical consequences. Whether we believe that high cholesterol diets have medical consequences or not, they do. The consequences affect some individuals more than others, but they do have bodily consequences. Whether we believe diets high in salt and sugar have clinical consequences or not, they do. The consequences affect some individuals more than others, but they do have health consequences sooner or later. Subsequently, if 90 percent of what we eat constitutes meat and animal products, our health is not going to be as optimal as when 90 percent of what we eat constitutes plants and plant products. Our ideas on dieting have consequences when they direct our eating choices.

Since this book is being written during the Covid-19 pandemic, a reference to scientific evidence coming from the World Health Organization (WHO) and the Centers for Disease Control and Prevention (CDC) must be noted. By checking with WHO and CDC, public decision makers and action takers try to garner sufficient and consistent scientific evidence pertaining to surviving the Covid-19 virus. When the scientific evidence is not sufficiently and consistently available, those who must decide and act for the public should do their best to recommend healthy and sane directions to take. Likewise, when we comment on vegan and non-vegan eating habits in this book, we do our best to launch our statements from sound scientific evidence. When the scientific findings have conflicts or inconsistencies, we do our best to note the limitations in our dietary recommendation. We will now uncover the flavorsome and healthful options that constitute the science-based practical vegan diet proposed in this book.

Part 1: The Green Light

While traveling in Hawaii, Richard Fiordo, one author of this book, encountered a "vegan" restaurant with a "vegan" menu. By omnivore standards, the vegan menu listed both savory and unsavory options typically found on such a menu: vegan imitations of hamburgers, hot dogs, ham, milk, eggs, and cheese. Due to the largely unappetizing menu at this Hawaiian vegan restaurant, the inspiration to explain and share tasty plant-based meals surfaced immediately and powerfully. After all, vegan meals can taste great and have tasted great for centuries. He thanked the restaurant *owner* for his inspiration and decided to write this book with Dr. Noah Chelliah—the other author of this book as well as a nutritionist and cardiologist who practices non-invasive, invasive, and interventional cardiology.

If you see this book as a sales pitch for the ecstasies of vegetables and the agonies of meats, you are not correct. This book is *not a vegan sales pitch*; it is *not an attempt to push you into veganism*. So *please do not stop reading* it if the idea of veganism is unsavory to you. Rather, this book is a *revelation of* "practical veganism" and *an invitation to you* to consider practical veganism for your dietary health and well-being. Practical veganism allows for meat and meat products to be a part of one's nourishment—as small a part as individually possible yet still a part. The goal of practical veganism is to be scientific, medical, and sensible about eating meat and animal products; it is to use medical measures for healthy eating—even when a portion of what we eat is meat. In a lighter vein and with wit, Dr. Chelliah tells us: *"There are no true vegans in this world because most plant-based foods do have some bugs in them!!"* In this book, we are using vegan as a term for convenience. We prefer the term *"plant-based wholefoods."*

There is a wonderful story about the Buddha. After enlightenment, the Buddha (meaning "one who is awake") passed a man on the road who was stunned by the Buddha's radiance and peacefulness. The man asked the Buddha what he was: "Are you a celestial being or a god? Are you a magician or a wizard? Are you a man?" The Buddha replied, "I am awake" (Kornfield, 1993,

p. ix). By comparison, on a meek and modest level, we would like those interested in diets to wake up to practical veganism: yes, to be awake to another eating lifestyle. We do not expect practical vegans to glow externally with radiance and peacefulness, although they often do, but we expect them to glow internally. Maybe some will even glow externally!

We promise to do our best, through this book, to awaken readers to the following: 1) what is meant by "vegan" and "practical vegan" eating lifestyles; 2)how everyone can *pursue* a practical and wholesome vegan diet; 3) how easy the practical vegan diet is to achieve in contrast to the vegan diet; 4) why practical veganism can be a medically safe end in itself; 5) how total veganism might be realized by more than a tiny percentage of people by using the stepping stone of a practical vegan diet; 6) how savory and traditionally established cuisine meals from around the world may contribute to increasing plant-based meals readily, healthfully, and tastefully; 7) how practical vegan diets measured by meat to plant consumption ratios are unequivocally more *attainable and sustainable* than total vegan diets; 8) how practical vegan diets can be far more practical for health for most people over time than total vegan diets; 9) how diets high in healthy plant-based foods can improve digestion, prevent constipation, eliminate bloating, *prevent* diseases, and increase energy; 10) health advice, suggestions, and recommendations based on scientific evidence over customs and practices driven by "mad" marketers; 11) practical vegan options under conditions where no vegan options are available; and, 12) *the complementary benefit of exercise* (or comparable physical labor or activities), modest or vigorous, in connection *with a practical vegan diet.*

Please be mindful of a special feature in this book. There are many myths connected with diets, vegan and non-vegan. We use the name "Nabal," a fool described in the Bible (1 Samuel, Chapter 25), to refer to myths, half-truths, lies, superstitions, nonsense, and other ideas promoted by foolish ones about foods - ideas not supported by scientific studies. Nabal should be thought of as the "food fool" and will serve to espouse falsehoods and follies about foods. Nabal's food fool ideas go something like this: "Diets schmyites! Eat what you like and don't worry about it. Life is short." As for the wisdom of Solomon in Proverbs 1:7 that warns us that "Only fools refuse to be taught," Nabal responds with "Wisdom schmisdom! I eat whatever I want to eat. No one is going to tell me what to eat!" To Nabal, we say with regret: "So be it and amen!" We also thank him for providing us with a comical judge who will allow us to laugh critically at "food foolish" ideas as we perambulate through the pages of this book celebrating practical veganism—while respecting veganism. We will do our best to explain ridiculous notions of food consumption through commentaries in the name of "Nabal." Yet, we will do our best to replace Nabal's fanciful dietary notions with food facts established through scientific testing and measurement.

Part 2: Medical Background of Dr. Chelliah

Who is Dr. Noah N. Chelliah and what has he done so far with his benign and fruitful career and life? After earning his medical degree, he completed a three-year residency in Internal Medicine. He then completed a two-year fellowship in cardiovascular diseases. In 1982, Dr. Chelliah arrived in Grand Forks, North Dakota to establish the Cardiac Catheterization Laboratory at Altru (formerly known as The United) Hospital. He performed the first coronary angiogram in Grand Forks. Since 1982, Dr. Chelliah has performed thousands of cardiac catheterizations or heart angiograms in Grand Forks over the last 30 years. As the equipment became available, he also added to his cardiology skills balloon procedures, stent procedures, rotational ablation devices (nicknamed the "roto-rooter" device), and pacemaker implantation procedures.

As the "science officer" of this book, Dr. Chelliah speaks truthfully and honestly yet caringly about meat-based diets versus wholefood plant-based diets. If the truth garnered from scientific findings upsets the reader, Dr. Chelliah is sorry and so is Fiordo. If the truth garnered from scientific findings uplifts the reader, Dr. Chelliah is happy and so is Fiordo. We know that *many people want to hear that whatever they are eating is fine, dandy, and healthy.* To those people, we can only frown and express our regrets. Many people resist information from science. We know this. Many people cannot overcome their eating habits. We know this too. To overcome any negative habit requires Herculean efforts. We respect the difficulty that comes with changing any negative habit, including a dietary habit.

Foods that at one time are seen as beneficial may later be determined to be unhealthy. *Drugs that at one time are seen as beneficial may later be rejected as being dangerous.* So, we have to stay up to date with emerging scientific studies that may or may not support our medicinal and dietary habits—everything from vegetables and meats to sweeteners. Most people know that peanuts are healthy for most people; yet, for someone with a peanut allergy, consuming peanuts can result in anaphylactic shock. Most people can consume crab, lobster, and shrimp; yet, an individual with a crustacean allergy can suffer gastroenteritis from consuming crab, lobster, or shrimp. Allergies are conspicuous prohibitions: for example, hives resulting from the consumption of chocolate and rashes resulting from the consumption of various juices. Contaminants in some foods and drinks add to the risk of whatever we ingest. We must do our best to discover what foods and drinks are actually or potentially harmful and avoid them. In truth, *just as medicine can hurt us, food and drink can hurt us too.*

Subsequently, Dr. Chelliah alerts us to the dangers in what we eat and drink. It is up to us to act on his advice. We are free to eat and drink today what we like despite any clinical problems we may face later in life. Dr. Chelliah is not a member of the food police and is not interested in any such function. In contrast to the cautionary medical advice from Dr. Chelliah, a dentist Fiordo knew quipped that his dental advice was not going to be respected. So, he declared to his patients: "Eat and drink what you like. Get a bunch of cavities and then I'll remove the cavities and get richer than I am now." The scientific truth is Dr. Chelliah's guiding light and benchmark. He serves as the reporter and interpreter of the medical and nutritional information that is based on science, research, data collection, and evidence. If readers require additional scientific information on nutrition from Dr. Chelliah, he can be reached at this email address: drchelliah@heartnd.com or either of these phone numbers: 701-780-2000 or 701-775-0157.

A theme that Dr. Chelliah *often* repeats is that *since experience matters, choose experience—experience in life, education, healthcare, medicine, and countless other domains.* Experience matters especially when the experience entails scientific findings that are not commercialized or politicized, material evidence that can be objectively observed and measured, sensible criticism of methods and findings, commonsense interpretations of observations and studies and judgments, and the wisdom to be open to correction when advancements or errors are identified. To be clear, we side with well-grounded, sane, tested, and proven experience. We invite you to do likewise in spite of our skeptical times where science, expertise, and experience are challenged with wisdom or folly.

Information provided about health, medications, nutrition, and foods comes from medical data that Dr. Chelliah stands behind. He genuinely cares about the health of his patients and the health of all people. As Dr. Chelliah has repeated numerous times over the decades, he wants to deliver the scientific truth about the healthfulness related to practical veganism and veganism: that is, the benefits of nutritious wholefood plant-based diets. Since there are dietary grounds backed by science for practical veganism and veganism, Dr. Chelliah as a cardiologist and dietary specialist provides the data and evidence behind plant-based healthy choices. For example, through practical veganism and veganism, we can predict a drop in bad cholesterol, a rise in good cholesterol, improvements in digestion, and many other benefits resulting from eating systematically and intelligently with an emphasis on plants over meats. Other advantages will be recognized as we proceed.

Currently, Dr. Chelliah describes his professional cardiology practice as being non-invasive, invasive, and interventional. He is educated and skilled in all types of procedures commonly done for patients with heart problems and has extensive experience in the management of all types of heart problems. He is a Clinical Professor of Medicine at the University of North Dakota Medical School and has published several articles in leading medical journals, including one in 1992 in the prestigious *New England Journal of Medicine*. He holds the following board certifications: Board Certified in Internal Medicine, Board Certified in Cardiology, Board Certified in Interventional Cardiology, and Board Certified in Forensic Medicine. He has also been elected as a Fellow of several national and international professional organizations: namely, a Fellow of the American College of Cardiology, a Fellow of the American College of Physicians, and a Fellow of the American College of Chest Physicians.

Furthermore, Dr. Chelliah is one of a handful of physicians in the US who also treats patients using science-based plant alternatives to meat as nutritional therapies with the power to prevent up to 85 percent of all chronic illnesses and even cure many of the illnesses without the use of conventional medications. He lectures in the US and Canada on preventing and even reversing heart disease, diabetes, and cancer. Because Dr. Chelliah has a passion to teach everyone about healthful eating, he started an FM radio station in Grand Forks, ND, USA which is dedicated to promoting good health. He does host a weekly radio show on his own radio station KEQQ FM every Monday at 5 PM central standard time. The radio show is broadcast live on Mondays. He answers questions from listeners during his live show at 1-877-347-0011 every Monday between 5 and 6 PM central. You can listen to his Otter Health show streaming live at www.keqqradio.com. He does two other shows on another radio station at 1310AM KNOX in Grand Forks on Mondays and Saturdays from 9-9:30AM every week. He also travels extensively to give presentations on health all over the US and Canada. One such health seminar is available on YouTube: https://www.youtube.com/watch?v=R83m2taTIeU&t=415s.

If you or your church group or social group is interested in having Dr. Chelliah present a health seminar, you can contact him by email at drchelliah@heartnd.com or by phone at 701-780-2000. To help patients with heart disease, Dr. Chelliah stresses the value of regular heart-healthy exercise and offers "Heart Healthy" cooking demos in the local grocery stores and in his Grand Forks office—The Heart Institute of North Dakota, Ltd. The conscientious Dr. Chelliah carefully cautions people on the potential limitations of medications: *"There is no such thing as a safe medication. All medications are unsafe, but they are useful when they are needed to treat diseases which are even more dangerous than the medications!"*

When Fiordo learned from cardiologist and nutritionist Dr. Chelliah of his dietary plan of practical veganism, he was floundering in a carnivore rut of excessive meat consumption with almost no plant food in his hypercarnivore diet. After receiving fundamental practical vegan dietary advice from Dr. Chelliah, in just one month, Fiordo was able to "speed shift" from a plant to meat (PM) ratio of 10 to 90 to a plant to meat ratio of 70 to 30. In other words, Dr. Chelliah's caring and scientific practical vegan lifestyle allowed Fiordo to add spinach, kale, carrots, peppers, cabbage, beans, lentils, spices, and many other vegetarian and vegan items to his diet. He was no longer stuck in a carnivore rut. Practical veganism, as advanced by Dr. Chelliah, showed him a tasty, sensible, healthy, and even spicey way out of the carnivorous gutter he was in. With many other practical vegans and vegans, he thanks Dr. Chelliah profoundly for the nutritional wisdom and method that freed him from a nearly total carnivore lifestyle. Fiordo had not succeeded in the past with a shift from meats to plants and would not have been able to move in a plant direction without Dr. Chelliah's insightful guidance.

Please note that Dr. Chelliah was not always a vegan. He used to be a carnivore and delighted in steaks, fish, and chicken. Dr. Chelliah *often recites* the story of how *he consumed a porter-house steak all by himself 20 years ago. He enjoyed consuming animal products at that time, but what changed his eating habits was science!* He knows the dietary discipline and struggles involved in moving away from consuming meat and toward consuming plants. Anyone interested in lowering meat in their eating habits can do so with the help of Dr. Chelliah's approach to practical veganism. His patients can testify to the effectiveness of his practical vegan lifestyle, food choices, and savory cooking options. One of Dr. Chelliah's patients went on his weight loss program about a year ago and has lost 100 pounds *without limiting his food intake*. Dr. Chelliah often tells us that trying to lose weight by decreasing food intake will not work in the long run. He tells us the best way to lose weight is to *increase food intake with the right types of food.* Fiordo testifies personally to what practical veganism allowed him to accomplish in his battle to increase the consumption of plants and decrease the consumption of meats. With confidence, Fiordo urges everyone to try practical veganism on and see how it fits. Fiordo and Chelliah are optimistic that anyone desiring to eat fewer meats and more plants will succeed. How far any individual can go in the direction of veganism depends on many factors including supportive circumstances, family backing, knowledge of veggies, ability to cook veggies with appetizing results, and personal motivation to avoid meats and approach plants.

Part 3: Overview

Practical veganism, or a healthy wholefoods plant-based diet, allows everyone to overcome the daily dining pressures and restrictions that surround us in predominantly meat-eating societies. While complete veganism *may constitute the ideal dietary lifestyle for health* (with studies on veganism and health being abundant), *complete veganism does have its challenges*. Still, statistically speaking, complete veganism has shown itself to be impractical in the US, Canada, and most of the carnivorous world. In practical veganism, we are interested in establishing lifelong ratios of plant eating to meat eating on medical grounds alone. Practical veganism (or PV) involves a preponderance, preeminence, and predominance of plant to animal ratios for a flexitarian lifestyle that celebrates consuming as many healthy plant-based wholefoods as possible: that is, as many foods from plants as carnivorous individuals can survive and thrive on, readily and happily, given their life circumstances and motivations. In a sense, we believe that eaters should err toward leniency in selecting foods from plants rather than foods from animals. We also believe that eaters should strive to limit the amount of food derived from animals. Thus, *maximize plant eating and minimize meat eating.*

The higher the plant to meat (PM) ratio is, the greater healthfulness you will likely attain. Based on percentages, while a 60:40 PM ratio is good, an 80:20 PM ratio is better. Given these ratios, the total occurrence of all food consumed equals 100 percent. The favorable occurrence of eating plant-based food in these two PM ratios is 60 and 80 (Conchie, 1975). Of course, the ratio can be higher: for instance, 90:10 or 95:5. Technically, the first term in the ratio is the antecedent and the second term the consequent (Burrell,1996). Accordingly, the higher the plant food antecedent and the lower the animal food consequent, the better for the health of the practical vegan. Higher antecedents plus lower consequents equal stronger health for the individual. Some stalwart individuals may be able to attain total veganism: that is, a pure diet of wholefoods that are plant-based. However, with reasonable commitment and desire, most of us can aspire to and acquire the higher ratios of practical veganism—one gloriously reachable goal of practical veganism being to increase the plant antecedent and decrease the meat consequent.

In short, we know we are asking our readers to consider and try a nutritional lifestyle that has not so far been tried, or even considered, nationally or globally. We believe you may not care to learn about or implement our scientifically oriented dietary viewpoints, procedures, and suggestions. We believe you may find some of them too strange or hard to incorporate into your life and therefore may resist them entirely. Nabal says: "You're right. I don't like the sound of eating veggies. I like to eat only what my grandmother and grandfather ate on the farm: eggs and potatoes in a frying pan greased with lots of healthy and delicious lard, bacon, pork, fried chicken, barbequed steaks, buttered popcorn, soda pop, beer, and lots of other goodies from the good old days. I say that we should either eat the way of back in the day or kick the cook to the curb. We ate plenty of meat at grandma's and grandpa's farm!"

Nonetheless, we ask you to forgive Nabal and put any all-or-nothing and either-or thinking aside. Rather than thinking either plant or meat choices, we support thinking of plant and meat choices in proportional amounts. Rather than thinking that every meal must be vegetarian or vegan or our health is doomed, we encourage thinking in terms of high ratios of plant to meat meals—that is, vegetable to animal ratios. We want to help you combat the tyranny of dietary *shoulds* and *should-nots* (Horney, 1950) that drown us in questionable advice about what to eat and how to eat it, especially as regards the should and should-not tensions between vegan, vegetarian, carnivore, and omnivore meals. In other words, should I eat meat at all? Should I not eat eggs? Should I not consume more meat for protein and calcium? Should I not consume fruits, vegetables, meats, meat products, candy, alcohol, soft drinks, nuts, seeds, and whatever else is edible?

Scientific research sources for health and nutrition are critical in making sensible food choices. "Ah," says Nabal, "scientists and researchers are just a bunch of over-educated showoffs. You can't get two of them to agree on whether or not I should eat an egg. What do they know about making really tasty food? Not much. Just ignore them as I do." Importantly, we acknowledge Nabal's cynicism and consternation that research on food and nutrition undergoes constant critical scrutiny and may face controversy concerning what constitutes nutritional facts (or reliable information) and sound interpretation of the nutritional facts. So, we proceed with open eyes and open minds in dealing with dietary data and the interpretation of the data so that our conclusions about what scientific research delivers is continually improving and kept up to date.

Along with many worthy nutritionists, we hold that a high plant diet is ordinarily healthier and far less harmful than a high meat diet. We are aware that total veganism has drawbacks, especially the risk of a deficiency of vitamin B12 (Craig, 2009). A high ratio, maybe an 80 to 20 ratio, of plant or vegetable to meat or animal products means that we need not rid ourselves from meat and meat products 100 percent of the time. While a total vegan diet may be best overall for the human body, *it does not mean that we can never consume animal products in small quantities and remain healthy*. Clearly, most highly carnivorous people have not switched to a total vegan diet. Regardless of health advantages, it is very difficult to get people to go on a radical vegan diet. In a sense, a pure vegan diet is impractical for most human beings in the US and most other countries. A practical vegan diet offers a workable and beneficial option to the "impractical" total vegan diet. However, medically there are degrees of difference, significant differences, between people who eat meat three times a day and those who eat meat three times a month. Medically, there are degrees of difference, significant differences, between people who eat veggies (healthy wholefoods that are plant-based) three times a day and those who eat veggies three times a month. Human bodies tend to be healthier as they increase their consumption of veggies and decrease their consumption of meats.

Because a total vegan diet is strenuous or impossible for most people to follow, Dr. Chelliah tells us that researchers attempted to study the effect of simply changing the ratio of meats consumed in favor of vegetables from a ratio of 2 to 1 (meats to vegetables or MP) to an MP ratio of 1 to 1. For example, researchers studied the results of a practical vegan diet on men with prostate cancer. They found that the doubling time of the prostate tumor in the practical vegan group (the half-vegan group) slowed from 21 months to 58 months. While the cancer kept growing, increasing the consumption of vegetables and decreasing the consumption of meats slowed down the expansion of the tumor. In comparison with another study that required men to go on a total vegan diet for a year, Dr. Dean Ornish published a study in which men with prostate cancer were able to reverse the progression of prostate cancer when they went on a pure vegan diet for 12 months. Thus, the ideal ratio for shrinking the prostate tumor is no meat and only vegetables—a meat to veggie ratio of zero to one (or 0:1). In summary, while the vegan diet may be ideal for prostate health, the practical vegan diet can achieve results close to it.

For now, it may be helpful to think of a practical vegan diet as one that violates a vegan diet as little as possible by utilizing individual willpower and within the limits of cultural circumstances (such as US Thanksgiving). Inarguably, a practical vegan diet violates a pure or total vegan diet but tries, metaphorically speaking, to reduce the violations to venial misdemeanors rather than mortal felonies! Of course, practical veganism lowers the bar from veganism, but not so low as to be harmful to health. Practical veganism strives to put the bar at a reachable level for everyone, reachable daily and for life—not for just a weekend or a month. Practical veganism is like regular exercise for anyone given the daily and weekly pressures of a job and the pressures of living in a modern society. Veganism is more like the rigorous training program of an Olympic athlete like Olympic sprinting champion Usain Bolt (aka "Lightning Bolt")—not safe or workable for life for everyone.

Practical veganism aims to avoid what Fiordo calls the "Devils Tower" problem. Devils Tower, Wyoming (Wikipedia.com, s.v. *Devils Tower*) is a butte in northeastern Wyoming. The Tower—the first United States national monument established on 24 September 1906 by President Theodore Roosevelt—stands 867 feet from base to summit at a height of 5,112 feet above sea level. The top of the butte tower is relatively flat with the sides creating a nearly 90-degree angle and a nearly straight drop off the top, not a gradual slope off the top. So, unless someone has the skill to skydive off the top, a fall from the top would be fatal. With veganism, there are steep sides like Devils Tower that create an either on or off the butte psychology. There is not a gradual sloping path for exiting Devil's Tower. When we fall off the vegan menu, the fall is swift, straight, and crushing. With practical veganism, we just roll a bit down a gentle slope and get back to a high ratio of plant to meat eating at the next meal. Veganism is close to an exclusive either-or situation while practical veganism is less exclusive and more inclusive. Practical veganism just makes the goal attainable by lowering the bar to a reasonable level of health for everyone on planet Earth. We achieve a practical vegan lifestyle that is healthy without the Devils Tower risks of falling for an impractical, and for most, an impossible vegan diet. Practical veganism reaches a sensible and healthful goal safely and routinely. Figuratively speaking, all-or-nothing vegan thinking seems to invite a subluxation in a vertebra or a sprain in an ankle.

Because it is impractical for all but approximately 5 percent (and conceivably less) of the human race in the US, Canada, Europe, Scandinavia, Australia, New Zealand, and the United Kingdom, veganism's *shoulds* and *should-nots* can become emotionally stressful and pragmatically impractical: that is, unworkable. Statistics on veganism and vegetarianism show increases in many countries over recent years. However, we cannot ignore that the statistical procedures for collecting the vegetarian and vegan data are questionable rather than foolproof. Still, what has been collected to date, faulty or not, reveals an approximate vegetarian average of around 10 percent and an approximate vegan average of around 5 percent (at best) of the populations of the countries cited above (Wikipedia.com, s.v. *Vegetarianism by Country*). In the US, for example, vegetarians may constitute five percent with vegans making up three percent. In the UK, vegetarians come to around 11 percent but vegans to only one percent. In Canada, vegetarians reach an estimated 9.4 percent while vegans emerge around 2.3 percent. And, these numbers are low in spite of the fact that more edible and gustatory meat substitutes have been developed and marketed in the past five years in addition to the centuries of tried-and-true delicious vegan and vegetarian cuisine dishes that have been consumed in diverse nations.

As a medical norm with tolerances to fluctuate between veganism and carnivorism in degree, practical veganism becomes the poster child for wholesome eating *for all*. Practical (partial or incomplete) veganism offers success that is within reason for resources and motivations, usable, valid, reliable, feasible, likely, probable, accomplishable, and practicable: in plain language, capable of being done. Impractical total veganism may work well optimistically for 5 percent of humanity; total vegetarianism may work well for optimistically 10 percent of humanity. Practical veganism may work well for all of humanity: no compunctions, no mandates, no remorse, no contrition, and no arm-twisting. Practicing practical veganism is like choosing to travel from Maine to Seattle by train. Traveling by train from coast to coast becomes the railway version of a plant-based norm. Occasionally, travelers may want to deviate from the track *for a while* and then return to their cross-country route. So, the practical vegan may get off the vegan train *for a while* and travel off for a time by car, truck, or motorcycle for a carnivorous excursion with the full intention of returning *after a brief break* to the train's main route.

A 10:90 plant to meat ratio in practical veganism is only acceptable as a starting point with successive and gradual increases toward a 70:30 or 90:10 ratio. As a yoga master from Pakistan once told Fiordo, when we exercise, we should train with medical safety, lifespan, and healthspan in mind so that we are able to live functionally with mobility, strength, flexibility, and endurance as we age. Training as an Olympic or professional athlete *risks crippling side effects*. To exercise moderately for 30 minutes three times a week would likely be safe for most of us. To exercise for an hour a day reasonably might be safe for comparatively robust people. To train with high intensity for five hours a day is *of questionable benefit* for most people but may be practicable for the statistically few elite athletes. In other words, we do not have to exercise rigorously for five hours a day to be healthy, and we do not have to eat a purely vegan diet to be healthy. Ergo, for most of us, modest but regular heart-healthy exercise and a practical vegan diet will advance our physical condition under normal and even severe pressures from work, family responsibilities, raising children, natural catastrophes, job loss, loss of friends, deaths of relatives, personal strivings and frustrations, and other tensions in contemporary US life. However, woebegone Nabal has a different spin on exercise. As he puts it: "The truth is living a long and healthy life isn't just about broccoli and walking 10,000 steps a day. I prefer to exercise from my couch. I lift my favorite beverage for the night in a mug from a table next to my couch, take a sip from the mug, and then lower it to the table. And, that's enough exercise for me for a week!"

We believe that an aside on public health may be of value to our readers. In the US and Canada, we recognize and acknowledge that our lives can become strained and burdensome. Living with encumbrances can be daily life for many of us. Hardships though need not drive us into smoking, alcohol consumption, opioid abuse, or other harmful health and existential choices. Any substance or food that is known to be deleterious to our lifespan or healthspan should be avoided, even if we are living and working under dire conditions.

Since it is unhealthy, we will use smoking as an illustration. If we smoke occasionally in small amounts, we are not likely to suffer much, biomedically. So, if we smoke one cigarette once a week, we will probably prevail medically over health risks. In contrast, if our normal daily indulgence entails five packs of cigarettes a day or more, we may eventually face unpleasant or tragic clinical consequences. If someone's house suffers serious damages from a windstorm, at the time of seeing the damage to the house, the homeowner might be emotionally wrought and choose to smoke two packs of cigarettes. The homeowner will likely recover from this temporary increase in nicotine—as long as the smoking does not continue indefinitely. Fiordo had a friend who, for years, chain-smoked cigarettes and drank beer excessively. His friend used to ask him: "Which habit do you think will take me?" In the case of the friend, he died of lung cancer from smoking. Again, a heavy consumption of cigarettes in terms of quantity and frequency can become a fundamental threat to health and life. *One* cigarette once a week or once a month would be recoverable for most of us. Five packs of cigarettes seven days a week would have people calling us "suicidal" and would likely add up eventually to health problems or death. Subsequently, medically speaking, worry more about smoking a lot daily rather than smoking a small amount on occasion: likewise with eating unhealthy foods on occasion rather than throughout the day! Nabal exclaims: "If you're right, I'm a goner. I begin my day with cigarettes, smoke on the job, and eat junk food. After work I shift to cigars and unhealthy liquids!"

Our goal is to help everyone learn how to become competent in identifying healthy and unhealthy vegan food choices, learn how to select healthy and tasty vegan and practical vegan choices, and learn how to become our own chef for delicious vegan or practical vegan meals. We borrow a few ideas from Dr. Claudio Violato (2019, pp. 37-38), a specialist in medical education and assessment. He is a Professor and Assistant Dean at the University of Minnesota's Medical School. He discusses competency based medical education (CBME). We take from this idea of CBME the idea of competency based dietary education (CBDE). Violato explains an emerging focus in medical education that deals with competency-based testing of "entrustable" professional activities (EPAs). From this test, we develop the notion of competency based self-testing and self-assessment of entrustable dietary activities (EDAs). EDAs pertain to identifying, cooking, and eating a practical vegan menu on our own. Of course, we have to be extremely careful with self-testing and self-assessment because they run risks of being inaccurate, for we can be unskilled and unaware of our decisions or over-evaluate our abilities (Violato, 2019, p. 121; Kahneman, 2011; and Kruger & Dunning, 1999).

Most of the time, we do not have an external tester or assessor whom we can rely on for the successful fulfillment of our dietary goals. Because most of us have to rely on our limited and biased self-assessment powers, we must be especially astute in seeing that we are on the proper practical vegan path. According to Violato, entrustability in medical education is achieved when a healthcare provider learns how to perform a professional activity in a safe and effective manner independently without a supervisor. Since the entrustable learner is then seen as being capable of functioning autonomously, responsibilities are turned over to that learner. We extend this worthy idea of entrustability from Violato to being competent in selecting, preparing, and cooking healthful and flavorful vegan and practical vegan meals self-reliantly without direct supervision. Practical vegans and vegans must get used to assessing their eating habits without having a dietary judge (Fiordo, 2018) over them. Each has to be a dietary judge of what should and should not be included in plant-oriented dining.

An illustration through dear old Nabal of an EDA seems to be in order. Nabal declares that the following two days of consuming what he calls practical vegan meals are healthy and commendable. Here is what Nabal the food fool plans to eat on Monday. Breakfast will be 3 eggs sunny side up with 3 pork sausages, 3 strips of bacon, bread coated with butter, and a glass of hot chocolate milk sweetened with sugar. Lunch will entail a double cheeseburger or chicken nuggets and a chocolate milkshake for the beverage. Dinner will include a 16-ounce prime rib steak with mashed potatoes with the skins included and sprinkled with bacon bits, and a pint of cola for the beverage. Between meals, Nabal will snack on beef jerky. His dining plans for Tuesday include the same thing except that between meals he plans to snack on a combination of carrots, cheese with raw tomatoes, and raw white mushrooms. Nabal has shown he is not entrustable with a vegan or practical vegan diet. As you learn what to select, prepare, and cook through what is presented in this book, you will see that Nabal is part of the madding crowd of crazy and sickening eaters. Unfortunately, the deluded Nabal wrongly believes that his Monday and Tuesday dietary plans are on the yellow brick road to healthy practical vegan and vegan dining.

Part 4: Source, Context, and Circumference of Practical Veganism

The term "practical veganism" should gain clarity, utility, and currency as we proceed. Practical veganism (PV) derives from Dr. Chelliah's medical and research experience. We acknowledge that we live at a time when just about anything asserted or explained can be, may be, and is challenged—whether on reasonable grounds or not. Hence, we expect that our readers may disagree with the lifestyle modification advocated in this book. However, Chelliah and Fiordo stand by the conclusions advanced in this text until additional scientific research compels us to change a particular view or position. We also know that some dietary assertions and suggestions will not be palatable to many people. While we do not wish to offend you with the research-based truth, we want to present the evidence to you for your nutritional benefit—like it or not. For the sake of improving everyone's health, we strive to stick as close to the nutritional facts as our powers and limits allow. Depending on the food or medicine studied, research supporting nutritional conclusions may vary from an abundance of studies that deliver results with high confidence to research with only a few studies with only modest confidence. Please note that we do our best to indicate whether the results of the research are based on a lot of studies or just a few studies. Nabal's reaction is this: "Well, you have failed. I'm offended. You've insulted me and stressed me out. Period."

In short, all research is not equal. Some areas of medical and nutritional research have received heavy attention and are mature while other areas are merely in the genesis of their investigation. We stress that many food and health studies deal with correlation (or dependency) between variables and not with causation. Two variables correlated do not necessarily equal causation. While the height of a father is correlated with the height of the son, the height of the father does not cause the height of the son. Underlying genetic inheritance laws influence the correlational findings. Because correlation only indicates a relationship, cause and effect cannot be interpreted from correlation. In addition to correlational data, to make inferences about cause and effect between two variables, experimental evidence must be provided. To draw causal conclusions from correlation risks being non-sensical (Violato, 2019).

The practical veganism generated and championed by Dr. Chelliah builds from the dietary premise that no one diet fits all. There is no one size fits all in dieting, and PV recognizes and works from this fact. If one diet did fit all humans, there would not be a plethora of diets available around the world. Practical veganism is a flexitarian diet that aims to be as wholefood plant-based as any one person can embrace. A healthy practical vegan diet offers the best possibilities for being the healthiest human being. For success, an early adoption in life of a PV eating lifestyle would be advantageous to the individual *if medical conditions support it.* While the PV diet can be started at any age, starting early in life, health conditions permitting, a predominantly plant-based wholefood diet would benefit any individual. Childhood and adolescence would be suitable times to initiate a PV eating plan. Practical veganism claims as worthy pursuits the enhancement of human lifespan and healthspan. *Practical veganism aims to have all of us live longer lives with a higher quality of life.*

The practical vegan centered diet provides ends and means for individual success in a healthy and mainly wholefood plant-based diet. Gradually improving an individual's plant to meat ratio toward a vegan orientation may be generally desired but not generally attainable. The PV diet allows for a progressive and gradual movement through the measurement of plant to meat ratios. If today, an individual has a plant to meat ratio of 10:90, a year from now, with dietary and culinary education, that individual can have a plant to meat ratio of 90:10. The PV diet proposed by Dr. Chelliah is based on research, evidence, data, and experience. Medical grounds serve as the critical measure for how much meat an individual's body can tolerate without serious negative side effects pertaining to such conditions as heart disease, cancer, diabetes, and obesity. The PV approach to meals does not affiliate itself with cultural extremes that avoid even trace amounts of meat or meat products. The PV diet aims to direct individuals to healthy meals on medical grounds. Vegan extremes are recognized yet not celebrated, given current medical research and support. The goals of PV include showing the importance of a person-centered health plan, spreading the word to the world about this healthy and nutritional plan, using careful and wise communication to spread the word, and nurturing individual choice as the best possible way to dietary health.

What guides us on our practical vegan healthy food journeys is how much meat our bodies can handle safely. How much meat can we consume without developing medical disorders? What are the facts? What does medical science have to say about a total vegan or high vegan diet? How much meat can we consume without facing health hazards? What is a generally safe ratio between plant eating and meat eating? Since no two of us are identical physically and medically, each of us has a different tolerance for meats and plants. You may be able to consume more or less meat than your closest friend without suffering cardiac or other health problems your dear friend may suffer. We are told that, for *"medical" reasons (which Dr. Chelliah disputes is a valid reason), even the Dalai Lama eats small amounts of meat.*

Two studies will be explained to emphasize the following ease and comfort connected with the advantages of implementing a practical vegan over a total vegan lifestyle. Given US and Canadian eating habits, the practical vegan diet is workable for just about everyone, if not everyone, while the total vegan or complete vegetarian diet faces the obstacles of omnivorous conditioning, dining circumstances, ignorance of savory vegan eating options, knowledge of only unsavory vegan eating options, and other personal and cultural antagonisms toward vegan and vegetarian eating.

A study by Hargreaves et al (2021) examined vegetarian diets through the perspective of quality of life. "Quality of life" in this study encompassed physical, psychological, social, and environmental realms; it also involves a subjective consciousness of purposefulness and wellbeing. Adopting a vegetarian diet has the potential to influence and be influenced positively or negatively by the physical, psychological, social, and environmental realms. Adopting a vegetarian diet could have positive results like better physical health and a lower impact on the environment. Vegetarianism though could also have a negative impact on the quality of life of those choosing to abstain from meats and other animal products, especially when vegetable availability is limited. Understanding the effects from the four realms is essential before adopting a vegetarian diet, not just its nutritional benefits. By comprehending and considering the many factors that might prevent people from adopting a vegetarian diet, studies of the quality of life might contribute to mitigating those factors that block or delay people from becoming vegetarians. Our interest in this PV book in explaining and promoting practical veganism provides one sensible and holistic way to lessen the physical, psychological, sociological, and environmental factors that can thwart eating for healthfulness more plants and fewer animals.

Another study by Katcher et al (2010) addressed a particular and concrete experiment with a worksite vegan nutrition program. The researchers noted that although vegetarian and vegan diets are effective in preventing and treating several chronic diseases, their acceptability outside a clinical trial setting is questionable and has not been extensively studied. The Katcher study was designed to determine the acceptability of a worksite vegan nutrition program and its consequences on healthy quality of life and work productivity. The researchers concluded that a worksite vegan nutrition program can be readily accepted and can contribute to the improvement of the health, quality of life, and work productivity of the employees. This study gives the practical vegan approach of this book support. If implementing a vegan nutrition program in a worksite showed success, we can rest with some data-based assurance that a practical vegan nutrition program might flourish in worksites in general and perhaps even in society in general. At least we are optimistic that the practical vegan lifestyle would be easier to implement than the more rigorous and punctilious vegan lifestyle.

The authors of this book believe that we alone can become vegans or practical vegans, but it is hard for us to do so alone. We may need a little help from our friends. Nabal cries out: "Friends? Who has friends? I don't. So, who in the world is going to help me?" Relax Monsieur Nabal and please consider the authors of this book your nutritional friends. If your desire does not entail becoming a total or pure vegan, we implore you to consider improving your lifespan and "healthspan" through practical veganism: that is, through herb appeal with a meat garnish or meat side dish—yes, by allowing meats proportionally to be the lesser side item with plants as the main meal. If you now eat a meat to plant ratio of 90% to 10% and want to reverse this ratio so that you eat a meat to plant ratio of 10% to 90%, practical veganism can be a first step toward accomplishing your goal—and accomplishing it easily and guiltlessly.

As Dr. Chelliah has told numerous audiences, if you eat a 20-ounce steak tonight, you will likely be alive tomorrow. However, if you eat a 20-ounce steak nightly, you may eventually develop heart disease. So, we stand by eating as many vegetables and berries and fruits and nuts as we like without excluding the occasional meat meal or, at least, "meat garnish" with a largely vegetarian meal. A practical vegan diet sees the word "eat" in the word "meat." A practical vegan eating lifestyle allows us to eat meat, but it allows it in degree—not a 20-ounce steak nightly! Occasional binging on meat may happen. When it does, we do not expect the bingers to give themselves 20 lashes with a whip. We simply advise that for several days after the meat binge, the binger either fast from meat and meat products or minimize meat and meat products in an amount that approaches total abstention. In short, we urge the binger to get back on the practical vegan track and in the practical vegan groove. That is all, folks.

Dr. Chelliah wants everyone to recognize that additional statistical and scientific studies can modify, or even reverse, conclusions as new studies are published. He compares and contrasts the DART I study in 1989 and the Dart II study in 2003 *authored by the same team of researchers.* Both studies will be detailed later in this book. Each study delivers different and conflicting information on the value of Omega-3 from fish and fish oils. Dr. Chelliah reminds readers to follow the money when it comes to nutrition and pills. Because Omega-3 from fish oils constitutes a billion-dollar industry, everyone is advised to be on guard about claims about the ultimate value of Omega-3 from fish oils. He emphasizes that Omega-3 can also be acquired from plant-based sources like algae at about the same price without the risks attached to fish oils. Yet, Nabal argues: "There's no difference. Omega-3 is Omega-3 and that's that! It's just much ado about nothing. I won't listen to anything else." However, Dr. Chelliah points out correctly that fish do not manufacture Omega-3. Fish acquire Omega-3 by consuming the algae in the water. In a comparative risk to reward ratio, the Omega-3 sources from fish oil run high risks for questionable rewards while the Omega-3 from algae run low risks for unquestionable rewards. Again, he does not catastrophize the consumption of Omega-3 from a bottle of fish oil capsules. There is no likely immediate fatal end from consuming a bottle of Omega-3 from fish oils. He just advises that we get the Omega-3 from plant sources, reduce health risks, and increase health benefits. In brief, Dr. Chelliah queries: *"Why get the Omega-3 through toxin-contaminated fish oil when the same Omega-3 can be obtained from golden Algae that the fish get it from!!"*

Flavorful and healthful vegan and practical vegan menus exist and more can be invented. We discuss flavorful vegan and practical vegan menus. Such menus emphasize to the hungry people of Mother Earth to seek wholesome, heart healthy, and savory vegan and practical vegan cuisine from around the world. Drawing from an Arby's commercial, Arby's may have the meat, but vegans have the meatless. Practical vegans, of course, have the meat and the meatless with the meatless over the meat. Again, the word "eat" can be a practical culinary guide, for it is a part of the word "meat" and "meatless." According to individual leanings, practical vegans may choose to eat smaller portions of meat as gauged by meatless options: for example, meat reduced to 40%, 20%, 10%, 5%, or even 1%. When in doubt, due to pressing circumstances at a restaurant or social gathering that features meaty items, a practical vegan may opt for a bowl of Chinese vegetable fried rice, Korean kimbap, Japanese sushi, Italian pasta with marinara sauce, Mexican vegetable tamale, French vegetable quiche, or simply fast for one meal! The PV has options and utilizes them as sagely as possible.

To an omnivore, being or becoming a total vegan can be taxing—like training to become an Olympian athlete. The road to veganism, for most human beings, is as strenuous as becoming an Olympic athlete. In contrast, to be a practical vegan is like exercising routinely outdoors or at a gym to augment fitness, flexibility, strength, endurance, and health. Olympian gold may not be present with the practical vegan because wellness goals may be front and center. If you will tolerate a wee bit of wordplay, we heard from the herd of interested dieters that they desired the primary worthy aim of this book to be *the encouragement of all people to improve their herb appeal.* In improving their herb appeal, optimistically they will improve their curve appeal and subsequently their herd appeal.

To repeat and expand, our approach to practical veganism relies on being *evidence based* and *conditions based.* Dr. Chelliah serves as the source of evidence-based research on foods and nutrition while both Fiordo and Chelliah serve as the sources for conditions-based information fulfilling culinary and eating choices for a practical vegan lifestyle. For medical and fitness purposes, we urge everyone to eat as many nutritious plant-based meals as practically possible so that they can enhance their health by preparing their meals with herb appeal. Whether omnivores have a goal of becoming strict or pure or radical vegans or not, *we urge everyone, given their circumstances, to eat as vegan as they practically* can and to do so without guilt as practical vegans.

Anyone and everyone can make an effort to increase their plant-based meals daily and weekly. Dr. Chelliah's brother who was also a physician used to say to Dr. Chelliah, "*good food must first get past my lips to do me any good*!" We applaud this saying. Whatever you like you will tend to eat. If you like garbanzo beans but do not care much for Lima beans, then eat garbanzo beans. If you love cabbage but do not care much for lettuce, eat cabbage. *Proper cooking technique is vital in making vegan meals tasteful*. How we cook the vegan dishes is critical to us liking the taste and looking forward to eating flavorsome meals in the future. Tantalizing cooking makes a huge difference. If proper spices, plant combinations, and cooking styles are utilized, the meals will be delectable and delightful for continued dining. Subsequently, we should eat what we like that is also wholesome and cook the food in a tasteful manner so that we stay on the practical vegan road with wide *finger-licking* smiles.

For healthy eating habits to form, we believe that eating what must first get past our lips is a reasonable plan and guide, and it can be achieved by adopting established succulent international vegan cuisine for wholesome meal choices. By "cuisine," we mean established cooking characterized by deliciously distinctive ingredients, techniques, and dishes that are usually associated with a specific culture, combinations of cultures, or geographic region (Wikipedia.com, s.v. *cuisine*). Cuisine dishes can be influenced by ingredients available locally. Meals highly suitable for a "practical" (or limited or partial) vegan lifestyle would include such cuisine favorites as spaghetti with marinara sauce (or marinara sauce with tiny portions of meat) or vegetable fried rice (or beef fried rice with tiny portions of meat). For practical vegan (vegan-like) eating, we can have the meat free pasta (or rice) on one occasion and the meat pasta (or rice) on another occasion. When no vegan options are available, a situation that may be encountered frequently when traveling, practical vegan options can be quite simple. When hungry and when in doubt about what can be done to quiet hunger when no vegan options are available, we can satisfy hunger temporarily by filling up on water. If we are strong enough and have the will, when vegan foods are not available, we can forego food for one meal by fasting. Most of us living in high caloric cultures can afford to lower our caloric intake by intermittent fasting.

a Human and animal exemplars of veganism

What deserves emphasis is that many athletes are vegans or vegetarians. A partial list of distinguished athletes who are vegans or vegetarians include: Olympic track star Carl Lewis, marathoner Fiona Oakes, Olympic weightlifter Kendrick Farris, rugby player Mirco Bergamasco, professional wrestlers Amy Dumas and Tyler Bate, triathlete Rip Esselstyn, mixed martial artists Luke Cummo and Nate Diaz, bobsledder Alexey Voyeboda, hockey player Georges Laraque, mountaineers Gerlinde Kaltenbrumer and Kuntal Joisher, strongman Patrick Baboumian, road racing cyclist Czeslaw Lang, skateboarder Ed Templeton, bodybuilder Jim Morris, basketball player Kyrie Irving, boxer David Haye, rock climber Stephanie Davis, runner Scott Jurek, dancer Tanya Kay, cyclist Molly Cameron, figure skaters Charlene Wong and Surya Bonaly, long-distance runner Paavo Nurmi, discus thrower Al Oerter, sprinter Leroy Burrell, hurdler and bobsledder Edwin Moses, and Olympic wrestler Chris Campbell.

Of course, the list of vegans, vegetarians, flexitarians, and practical vegans goes on and on. And, for those who still fear a shortage of protein if meat is not consumed, we not only have human athletes as support for veganism and sufficient protein, but we also have wildlife as support. Wolves, bears, tigers, dogs, cats, and many of our other animal friends eat meat. However, elephants, gorillas, rhinos, buffalo, moose, elk, and many of our other animal friends do not eat meat. Yet, human athletes and very large wild animals live well feeding on plant life alone. The plant protein suffices to allow them to survive and thrive. Nabal protests: "I'm sorry, but I can't believe it. I need my protein and only meat can give it to me. Now, I'm hungry for a double cheeseburger."

b Practical veganism as a reachable goal for everyone

Through practical veganism, we wish to bring hope to the hopeless and help to the helpless omnivores of the world—those struggling to ground themselves in a plant-based diet while stuck in the typically mindless trap of either-or, all or nothing, thinking about being total vegans versus total carnivores or omnivores. We hold that progressing from a diet that has 90 percent carnivore meals to 10 percent carnivore meals will improve overall health and wellbeing. Subsequently, we work actively and reasonably from the dietary bias of a practical vegan, not the dietary bias a radical carnivore or radical vegan. As practical vegans, we accept the premise of falling from the grace of veganism. We do not fall from the grace of eating meat. Our goal is to minimize meat products; and, when possible, for a time, eat only plant-based foods. Our gold standard is that of the vegan, not the carnivore or omnivore. Yet, we realize that some people will have personal reasons which prevent them from following a full vegan lifestyle. As a reminder, we are told the living Dalai Lama, for personal reasons, consumes portions of meat, even though he would prefer to be a total vegan. *Dr. Chelliah thinks that the science behind this logic is flawed and based on faulty information.*

Practical veganism aims medically to lessen *inflammaging*—that is, "accelerated and deleterious aging driven by inflammation." Inflammation may increase as we age. However, our modern lifestyles increase inflammation. Modern lifestyles entail unhealthy food choices, little or no exercise, societal stresses, alcohol consumption, and exposure to pollution and other toxins. Consequently, we may suffer from chronic inflammation that contributes to age-related health decline and disease. Inflammaging may reduce our lifespan and *healthspan*—that is, the "period of life in which we are healthy and thriving" (Mueller, 2021). Without a doubt, we may sanely desire both a robust lifespan and healthspan. We may more likely extend both our lifespan and healthspan through a practical vegan culinary lifestyle. To live to be 99 years of age would be especially wonderful if we could be riding horses, cycling, chopping wood, and skiing the day before our end. In short, we endorse the notion that one's lifespan and healthspan optimally should result in a photo finish. Linking lifespan to "healthspan" is important since it underscores the practical benefits of an increased quality of life through practical veganism; that is, the longest, healthiest, and fullest life embodies the ideal we enshrine.

We link practical veganism to the notion of a disruptive innovation as explained by Christensen, Grossman, and Hwang (2009) in their book entitled *The Innovator's Prescription: A Disruptive Solution for Health Care.* What they say about disruptive innovation in health care, we say about disruptive innovation in a heart healthy diet. As a nation, we are in big trouble thanks to our terminally ill eating habits. Practical veganism draws from the inspiration of Christensen, Grossman, and Hwang's *disruptive innovation.* Disruptive innovation transforms the unattainable into the attainable, the unaffordable into the affordable, and the complicated into the simple. Practical veganism is not only for the elite athletes of Olympian fame; it is for everyone. "Disruptive innovation" in the sense used by these three authors "makes things simpler and more affordable" (p. 1). While a practical vegan diet may not be as ideal as a vegan diet, from a health science point of view, it can be good enough: that is, health will improve. We plainly aim to transform eating lifestyles that are complicated and expensive into those that are affordable, simple, workable, effective, tasty, and—when desired—spicy.

Commercial *weight loss* diet *programs promoted by* companies like Jenny Craig and Nutrisystem are costly, even when discounts are in operation, for the overwhelming majority of people; their cost ranges from approximately $10 per day to $15 or more per day. At $10 per day, thirty days on such plans can cost $300 and 300 days can cost $3,000. The likelihood is that the cost per year will exceed $3,000 is high. The commercial companies rely on an elite group who can or are willing to pay $3,000 a year (or more) to reduce their high caloric eating habits. For the rest of humanity, the cost is unaffordable, prohibitive, or undesirable. In short, these diet food factories do not constitute disruptive innovations in dieting. They are like the costly Tesla electric vehicle, affordable to a select minority. When these diet factories and Tesla make products that are affordable and desirable to millions of people, a disruptive innovation in these industries may occur. In the meantime, both remain outside the economic range for most of us. Practical veganism brings to our eating lifestyles a plan that is simple and affordable. Therefore, practical veganism rides proudly on the powerful back of a disruptive innovation. Dr. Chelliah states: *"If any of these companies is truly successful in its mission to cure obesity, it should go out of business in a few years. Perhaps the reason why it is still in business is because its program does not work long term!!"*

While total veganism is attainable by some, practical veganism is attainable by all. In a few words, total veganism is impractical. We aim only to specify a scientific and medical basis for a vegan-like diet, not a religious or political or metaphysical basis. We ask these medical, scientific, and material questions: How much meat can be consumed without hurting our bodies? How many herbs can be consumed without hurting one's health? What are the medical consequences of eating mostly animal products daily versus eating mostly plants daily? Is a diet high in herbs healthier overall than a diet high in meats? Again, the scientific findings guide us in answering these medical questions in favor of vegetation.

c Veganism and related terms and concepts

The clarification of key words comes next. Admittedly, the terms that follow are simplifications from the many words used to describe extremely complex and diverse eating habits, such as: fruitarian, macrobiotic, and pescatarian diets. As various nutritional authorities inform us, human beings tend to be omnivores, and omnivores eat herbs, meats, and other morsels. *Omnivores* may range from being almost exclusively *herbivorous* (that is, hypocarnivores or hyperherbivores) to almost exclusively *carnivorous* (that is, hypercarnivores or hypoherbivores). In the current text, *omnivores* refer to practically attainable eating options: for example, everything from organically raised spinach and grass-fed beef to "junk-foods" like "cotton candy" available in conventional grocery stores. *Omnivore* here does not include such exotic items as fried grasshoppers, locusts, or worms. *Vegetarianism* entails the practice of abstaining from the consumption of meat (red meat, poultry, seafood, and the flesh of any other animal); it may include abstention from by-products of animals, such as dairy and eggs.

Semi-vegetarianism refers to diets that consist largely of vegetarian foods but may include fish or poultry, or sometimes other meats, on an infrequent basis. Semi-vegetarian diets, also called *flexitarian* diets, center on plant foods with the occasional inclusion of meat. *Flexitarian* is a portmanteau of the words *flexible* and *vegetarian* and signifies a less strict diet pattern when compared to strict vegan and vegetarian diets: thus, meals that may integrate a small percentage of meat or animal products. While vegetarian diets strictly abstain from consuming meat and vegan diets abstain from the use of all animal products, flexitarian diets blend omnivore and carnivore choices with plant-based meals. *Veganism* is a dietary challenge that excludes all animal flesh and by-products, such as milk, honey, cheese, and eggs as well as items refined or manufactured through any such product, such as animal-tested baking soda or white sugar refined with bone char. *Raw veganism* includes only fresh and uncooked fruits, nuts, seeds, and vegetables. With raw veganism, food must not be heated above 118 °F (48 °C) to be considered "raw." Usually, raw vegan food is only "cooked" with a food dehydrator at low temperatures (Wikipedia.com, s.v. *vegetarianism*).

Practical vegans should prioritize their violations of vegan ideals. Not all violations of veganism are equal. Quantity and quality count in gauging vegan violations. Practical vegans eat animal products as sparingly as their powers allow, yet they eat animal products—again preferably on at least an 80:20 ratio of plants to animals. So, when meat is consumed, what might the priority be for eating meats in 2022 given the contaminants in the land, sea, and air? Assuming small amounts of meats are consumed rather than 20 ounces of beef, chicken, smelts, or shrimp at one sitting, how might science guide us in ranking meats, fish, and fowl? How might science rank a list of meats from least to most harmful? How much controversy might be a part of the ranking system? The following rank ordering from least to most harmful might be acceptable to a number of nutrition experts: wild animals (elk, deer, moose, musk ox, buffalo), lamb and goat, beef, fish and crustaceans, and chicken. In this order of dangerousness, Dr. Chelliah strongly recommends against eating pork, fish (including all shellfish, eels, and sharks), and chicken. Although camel and horse meat are not ordinarily available in the US and Canada, we should also avoid consuming meat from these animals. As for meats that are the least dangerous (yet not safe nor good) to consume are lamb, goat, elk, deer, moose, musk ox, buffalo, and beef. Again, chicken and fish are violations as healthy menu items. Thus, radical vegans eat no meat, eggs, dairy products, and any other animal-derived substances. *Practical vegans* practice semi-vegetarian or flexitarian eating habits with a strong leaning toward hypocarnivore and hypeherbivore eating practices: that is, as little meat as possible during the dining day and as many plants as possible during the dining day. Subsequently, practical veganism becomes a more *attainable and sustainable* goal for most people who want to reduce meat and animal-derived substances.

To reiterate, *veganism* here refers to totally plant-based meals free of animal products. The veganism we favor is a healthy vegan diet free of consumables like vodka, tequila, and many so-called "junk food" meals that are not healthy yet are plant-based. The practical veganism we highly endorse is grounded in the fact that our bodies can assimilate some animal meals with minimal harm; however, highly carnivorous meals of beef, fish, or fowl as well as products of animals like butter, cheese, milk, and eggs should be minimized.

While practical vegans do not have to meet the dining challenge of pure vegans, they should strive to minimize meat, cheese, eggs, or milk. The ratio of meat-oriented meals to plant-oriented meals constitutes the thrust of the practical vegan dietary mission and choices. We would applaud anyone striving towards a total vegan diet. We believe they would be more successful if they took their journey to veganism in *gradual, cumulative, and progressive* steps. A workable plant to meat ratio of 80:20 can function as a healthy operating goal for practical veganism. A digestible ratio for most practical vegan diets would be 80 percent plant meals and 20 percent meat meals, calculated per meal or per day or per week. A practical vegan and marathon runner from Kenya whom Fiordo spoke with, reported that if he has a large meat meal one day, he abstains from meat for one to three days afterwards. Sometimes he balances over several days what he gauges to be his 75 to 25 plant-meat practical vegan ratio.

Many commercial meals blend plants with meats in modest proportions. For example, chicken fried rice may have a plant to meat ratio of 95-5. Although practical veganism tolerates animal and products of animals in their meals, the ratio for a healthy body should be a comparatively low consumption of meat and its products and a high consumption of plants and their products, such as plant-based cheeses, milks, and eggs. As noted earlier, a practical vegan diet might also be called semi-vegetarian or flexitarian. We understand and tolerate, but do not celebrate, fanatical veganism and its social consequences. If individuals choose to be fastidious vegans, we applaud them while expecting most people to fall short of the demanding dietary standard and discipline of ultimate veganism.

Veganism and agnosticism must be considered. Many omnivores take refuge in the belief that there are no sufficient research findings to know whether veganism has advantages over omnivorous and carnivorous diets. With Nabal, they believe that enough is not known about the effects of vegan or practical vegan diets over omnivorous diets. They might argue that the research on veganism is inconclusive and confusing because *they do not know about the effects of vegan or practical vegan diets*. The agnostic argument may have grounds in metaphysics and religion, but it does not enjoy such an intellectual status in the science of nutrition and diet. The omnivore agnostics assert that there is nothing to hold onto with respect to the research supporting veganism. They maintain that research is mixed with as many benefits as drawbacks. They do not believe there is a preponderance of evidence that supports the benefits of a healthy vegan diet over an omnivore diet. Nabal agrees wholeheartedly and mindfully with them. However, Dr. Chelliah knows Nabal is wrong and maintains that the scientific research findings on veganism show that there are clear advantages of a healthy vegan or practical vegan diet over an omnivore or carnivore diet.

Being semi-vegetarian and flexitarian, a wholesome practical vegan diet may lead to a diet high in veganism or to total veganism. The goal of a practical vegan diet may be nothing more than a healthfully satisfying plant to meat ratio of 70:30 or it may be total (or pure) veganism. It may also provide a clear basis for consuming a preponderance of plant-based foods over animal-based foods—again, perhaps a ratio of 90:10 or even higher, a highly dominant ratio of plant to animal. A practical vegan diet is thus progressive and partial in that it moves toward high or pure veganism; it is gradual in that it moves in degree upward in its plant to animal ratio: for instance, from 10:90 to 90:10. Practical veganism does not celebrate having a 20-ounce steak at one meal, much less a 72-ounce steak at one meal; rather, it may applaud progressively having 20 ounces of meat over five meals or more or 20 ounces this week, 15 ounces next week, 10 ounces the proceeding week, and so on. The quantity and distribution of animal products must be taken into account when judging the practicality, usability, and healthfulness of the practical (or partial) vegan way of eating.

A scale of plant to meat (PM) ratios might clarify relevant distinctions. Rather than deal with ratios based on hundredths, we will employ ratios based on tens. So, even though we acknowledge that a ratio of 1 to 99 or 22 to 78 are possible and even likely, we will stay with tens to keep the scale workable and plain: that is, 10 to 90, 50 to 50, 90 to 10, and so on.

<u>Scale of Plant to Meat Ratios</u>

10 to 90/10:90

20 to 80/20:80

30 to 70/30:70

40 to 60/40:60

50 to 50/50:50

60 to 40/60:40

70 to 30/70:30

80 to 20/80:20

90 to 10/90:10

d A social caution with respect to extreme vegans

A dietary and social caution may be helpful. Total, radical, and even fanatical veganism may create numerous obstacles to reducing meat and omnivore consumption for omnivores in the US and elsewhere. Unhealthy vegan choices may result in obese bodies instead of lean bodies. Vegans with a sense of humor may joke about their vegan choices including vodka, tequila, wine, beer, sodas, and other less than healthy beverages as well as candies and snack foods (such as French fries) that are other than healthy. Furthermore, extremist vegans can put veganism at risk by spawning boomerang effects with fewer rather than more converts to veganism.

Although we respect enormously and sincerely those who strive to eliminate even trace amounts of animal products from their meals due to religious obligations or moral sensitivities related to killing animals, we are concerned in this book only with scientific, medical, and physiological grounds for maintaining a healthy plant-oriented diet. Chelliah will detail the scientific, medical, and physiological conditions under which a practical vegan chooses nutrition. For example, forgetting that elephants and gorillas are vegans, many human omnivores and carnivores think that a vegan diet will not provide sufficient protein. On the other hand, some radical vegans might believe that having a huge steak with a large chocolate shake is not only immoral but will cause a vulnerable diner a heart attack on the spot. Nabal recognizes this may be a problem for some people, but not for him. Later on, Nabal may have a 72-ounce steak with lobster and shrimp washed down with a liter of beer. Chelliah will explain the err in his thinking based on scientific findings, not imagined or subjective health expectations.

Jokes are told about vegans who approach fanaticism. Out of respect for their presumed devotion to the cause of veganism, we will not tell any here. However, we will relate two stories from experience about two decent human beings Fiordo knew. One of these decent human beings was a kindly vegan extremist. While dining at a restaurant in Billings, Montana the vegan extremist caused a fracas and bruhaha unintentionally because a trace amount of a meat product was suspected in the lunch ordered. This kindly yet inflexible vegan created such chaos in the restaurant that the owner of the restaurant gave this sincerely radical soul money to go to another restaurant to satisfy an obsessive vegan hunger that could not be satisfied at the owner's restaurant—a restaurant that had food products with trace elements of animal products. A less provocative vegan approach to dinner would have been to order a fruit or vegetable dish—or, more radically, to drink water and not consume solid food at all! A second undiplomatic tale involved a rather transgressive extremist vegan. To provoke, insult, belittle, offend, and embarrass carnivores, another radical vegan Fiordo knew would routinely ask omnivores and carnivores: "Why are you still devouring flesh?" We grasp the discomfort and frustration of such a hostile statement from a fanatical vegan, yet we believe the hostility can be expressed with more diplomacy.

Not all omnivores have the Olympian discipline, desire, or vegan food networks available to become pure vegans. However, to eat a semi-vegetarian, highly veganized diet can be achieved by all omnivores choosing to increase their plant-based meals and decrease their animal-based meals without diplomatic disasters. Radical vegans are respected in this book, but we neither sanction nor advocate fanatical vegan diets for the billions who are omnivores. We do sanction and advocate, however, practical veganism for everyone on planet Earth. Please underline the fact that a radical vegan need not be a fanatical vegan. A vegan who does not want to consume even trace amounts of animal products may proceed with wisdom, gentility, and reasonableness when dining at non-vegan restaurants. Some total vegans, rather than creating a donnybrook at a non-vegan restaurant might unassumingly order a simple dish of French fries, especially those prepared in an air fryer. Some may choose to fast for that meal and settle for water only.

Truly, a vegan oriented diet can be challenging in the US, Canada, and other countries that consume many meats and animal products. Consequently, we focus on foods, vegetables and meats that are available almost everywhere in the US and Canada and suitable for practical vegan dining. Rare vegetables like shallot and moringa as well as rare meats like deep fried alligator and Cantonese snake soup are typically not available in the US and Canada. In different parts of the world, many foods and delicacies favored outside the US and Canada would not likely be considered for common consumption in the US and Canada—such as, blood from goats and cattle as well as meat from dogs and horses. Chelliah will show how the practical vegan challenge can be met on practical, scientific, and evidence-based dietary grounds. Scientific studies cited by Chelliah will provide justification for healthy practical vegan and vegan foods. Remember that he is the science officer of this book on practical veganism and the originating guru for a "practical vegan" eating lifestyle. Subsequently, his assertion prevails: Experience matters, so choose experience—especially Dr. Chelliah's *extensive* medical and nutritional experience.

e The 80:20 practical vegan ratio magnified

For people desiring to reduce gradually and progressively their meat consumption (or, more broadly, animal-based foods) and perhaps even eliminate meat consumption, a practical vegan diet can be instrumental in achieving that goal since it allows for gradual improvements. The practical vegan eating plan may spare many from what psychiatrist Dr. Karen Horney calls the neurotic and counterproductive "tyranny of the *should's*": obsessive and compulsive demands placed on oneself to maintain a perfectly pure vegan diet, no matter how demanding and discouraging the total vegan diet may be for a supermajority of human beings. Most human bodies can eat small amounts of meat and stay healthy. As a practical measure, to move from a plant-meat ratio of 20:80 to a plant-meat ratio of 80:20 usually results in improvement in nutrition, digestion, and health.

For those with a practical vegan orientation, eating even 70 percent vegan meals upgrades general health and well-being, notably when the practical vegan diet is also nutritional and natural. Note though that for weight control, the total quantity of food must be taken into account as well as the 80:20 ratio. If we eat an 80:20 ratio of plant to meat but consume enormous quantities of food, we will—without shock or surprise—likely add pounds. We cannot eat all we want even with the 80:20 ratio, or we will risk gaining weight. While the 80:20 plan is sensible, eating as if we are at an all-you-can-eat restaurant is not sensible if we want to maintain or lose weight. We have to push ourselves away from the table while respecting the 80:20 plan! For those who are new to vegan or vegetarian diets, in case you do not know this, be aware that most people can gain weight on a vegan or vegetarian diet. Inarguably, we should not overindulge on plant-based foods even if we are total vegans. We can gain weight from a vegan diet if the selections are not modestly sized and wisely chosen. Therefore, respect the 80:20 practical vegan plan along with calculated caloric amounts so that overweight is not a result! Nabal breaks his silence on this issue: "I don't want to gain weight. So, I'll avoid all vegan food plans so that I don't gain weight. Obesity would be hazardous to my health. I'll avoid veggies and stay healthy and lean."

With practical veganism, the daily plan for a plant to meat ratio may benefit from an explanation for practically minded eaters. Whether we assume for the supermajority of foodies a workable plant to meat ratio of 80:20 or a radically rare plant to meat ratio of 99:1, in our real-life culinary activities, we might follow a daily dietary plan something like this. Ideally, to dine by a doable 80:20 plant to meat ratio, the plant portion should entail healthy vegan choices. Healthy vegan choices would exclude vegan alcoholic beverages, candy, junk foods, and the like. Healthy vegan choices would include vegetables, rice, quinoa, fruits, nuts, berries, and vegan cheese. Acceptable common meat choices in the US and Canada might include grass fed beef, lamb, and carefully prepared wild meats from deer, elk, and moose. Unacceptable meat choices would include chicken, fish, and pork—all high in toxic ingredients. Daily, we should measure and weigh the vegan options against the meat options to make sure our total vegan and meat foods amount to the 80:20 ratio. So, 80 percent of your food for the day should be plant based and only 20 percent meat based. However, this ideal is based on "healthy" choices. In real life, we may fall short on both plants and meats for the ideal of healthy choices. Most of us though will survive this blunder!

So, what next? Is all lost? Absolutely not. Nonetheless, Nabal shouts from a hilltop: "All is lost. Just cast your fate to the wind and eat whatever your little old and clogged heart desires!" In an imperfect world with imperfect dieters, we must expect to work under imperfect circumstances; that is, circumstances that fall short of the glory of hearty culinary options. Life can be demanding, and few of us are being supported to train for Olympic competition. When we are stressed, we may turn to inferior foods—even harmful foods. Practical veganism tolerates us resorting to junk foods on occasion, but then returning immediately to the track of healthy foods. Clearly, we encourage practical vegans to opt for healthy food choices whenever possible. When optimal choices are not selected due to limitations in available foods or to limitations born out of the strained or yummy mood of the eater, we may fall from the grace of healthy choices. Our fall though should be rare and not routine. As a forgivable deviation from wholesome 80:20 food choices, we should return to our track of nutritious choices pronto.

For example, on one day, 80 percent of the daily food allotment might be healthy, robust, and premier vegan while 20 percent of the daily food allotment might be unhealthy, feeble, and lowly meat—such as a can of corned beef hash with three eggs and a pint of whole milk. We should worry not, for the weakness for this day can be strengthened the next day. Our biology will allow for the pathetic 20 percent meat meal on this day with a highly nutritious vegan meal the next day. Let us not forget that many human beings routinely go to bed hungry because food is neither available nor affordable. Under such dire conditions, one bad meal can be better than no meal at all. Ask a hungry soldier on a battlefield or a cowpoke on a cattle drive! The formula might be displayed through these four categories:

I	80% = health-giving with 20% health-giving
II	80% = health-giving with 20% health-taking
III	80% = health-taking with 20% health-giving
IV	80% health-taking with 20% health-taking

Indubitably, we should strive toward Category I and avoid Category IV! When we go on an eating binge, so be it. We may return to our Category I 80:20 regimen the next day with minimal medical stress and no guilt. If the eating binge lasts a month, the medical damage may still be recoverable, but the risk is not worth our gluttony. Amen.

The 80:20 ratio tolerates less than ideal choices as a rule. We should not flagellate ourselves over faulty or less than quintessential dining decisions. We should simply get back in the culinary Shangri-la groove. Here is how to fulfill the ratio: Whether completely wholesome or not, our ambition should be to stick to the 80:20 ratio. Daily, our vegan portion of the 80:20 ratio might include cereal with non-dairy oat milk and a banana for breakfast, a meatless burger with vegan cheese for lunch with fruit and nuts along with filtered water and tea, and four ounces of sardines with rice and beans for dinner. Nabal's reaction follows: "A meatless burger sounds nauseating to me. Why would I choose to make myself sick? Yuk! Just give me a good old-fashioned double cheeseburger wrapped in bacon with plenty of salt, pepper, and catsup on it."

Please note this less healthy alternative. The meat part of the 80:20 ratio might be a 10-ounce kiełbasa wiejska (farmhouse sausage) with an egg, a large bun, and a glass of chocolate milk with plenty of sugar. The meat portion may deteriorate on some days into even unacceptable meat sources like commercial chickenburgers, raw tomatoes, processed cheese, and a thick chocolate shake. Regarding meat choices, if fitting and proper healthy eating opportunities are not available or if we choose to eat like a junk food rogue on the 20 percent part of the 80:20 ratio, stranger things happen. How can we be so cavalier about the naughty 20 percent selection? The debilitating food choice can occasionally be tolerated, biologically and medically speaking. Since it only counts for 20 percent of the meal, we will recover from a wild and wooly nutritional violation. No guilt is required. However, on the next day, we should try our best to consume wholesome choices of *more* plants and *less* meats. Sometimes, like the Kenyan runner mentioned earlier, we may choose to have no meat for several days. If over a period of perhaps five days we have zero meat for four days and 50 percent meat for the fifth day, we may still approximate an 80:20 plant to meat ratio. Some flexitarian practical vegans calculate their meat consumption over the span of a week rather than over the span of a day and still maintain a hearty diet. Although we do not approve of the ensuing style of eating, we will describe it. Some practical vegans may have zero meat one day and fried chicken the next day. They may have zero meat the second day and a double cheeseburger the next. They choose to recover from their meat consumption on one day by a meat fast the next day. Therefore, we can make reasonable eating choices without resorting to inane and unhealthy self-punishments. Science and medical health guides us, not cultural ideologies.

Part 5: Practical Veganism and Exercise

Of course, to be as safe and healthful as possible, *we advise that any planned changes in diet or exercise be presented to and approved by a physician.* As a cautionary procedure, if the reader has any risky medical condition, is taking medication, is pregnant or lactating, has an autoimmune condition, suffers from any physical or mental disease or disorder, is at an age that has dangers for dieting or exercising, *we recommend seeing a physician before changing your diet or beginning an exercise program*—aerobic, anaerobic, or a combination. Please *play it safe by obtaining medical approval first.* Once medical approval is acquired, we endorse a healthful exercise program ranging from modest to rigorous exercise three or more times per week. To have success with a practical vegan lifestyle, it is critical to integrate exercise into the PV dietary life. None of us should underestimate *the value of practical veganism plus exercise.*

As for eating and exercising, to advance health, we advocate supplementing our practical vegan dining habits by minimizing sugar and salt intake. For those who wish to progress to total veganism or who already are total vegans, we recommend Vitamin B-12 as a supplement to the meat that is not consumed in a radical vegan diet. We also urge those, *upon medical approval*, to perform *moderate* aerobic exercise five days per week and *moderate* anaerobic exercise three times per week. Aerobic exercise would involve walking, cycling, swimming, dancing, cross-country skiing, skating, jazzercise, shadow boxing, bag punching, and other oxygen demanding activities. Anaerobic (or resistance exercise) would entail such acts as lifting weights, calisthenics, isometrics, static contractions, and power yoga. Nevertheless, Dr. Chelliah warns us that weightlifting should not involve very large weights (like bench pressing 200 pounds or more when this amount of weight demands severe straining to complete it) that are hard on the heart. Nabal renders his decision on exercise: "I walk from my kitchen sink to the couch. That's far enough for me. I carry out the garbage. That is enough exercise for me. Besides, I firmly believe in my family tradition. They say that exercise wears out the body."

Assuming medical approval has been sought and acquired before launching an exercise program, we suggest a modest plan of anaerobic or resistance exercise. Since weight training is generally available at gyms everywhere, weight training would constitute a primary anaerobic focus with isometrics as a secondary anaerobic focus. Whether familiar or unfamiliar with weight training exercises (light or heavy) or with isometric exercises (overcoming or yielding forms), *we encourage you to select an exercise routine that you and your physician believes would be a safe and effective plan.* We ask you to consider practicing a total body workout that would stimulate muscles in the chest and back, thighs and calves, and abdominals (Safrit & Wood, 1995; Thomas, Nelson, & Silverman, 2005).

We approve and celebrate the anaerobic exercise regimen explained and promoted by Bob "Robert Steven" Gajda, a former Mr. America and retired coach, physical therapist, and personal trainer (Wikipedia.com, s.v. *PHA training*). The system he popularized from the theory of Arthur H. Steinhaus' Theory of Peripheral Heart Action (PHA) training is a form of bodybuilding circuit training. PHA Training requires the exerciser to exercise continuously. It is typically performed using multi-muscle exercises such as squats, pulls, and presses. PHA Training follows a specific pattern so that no one muscle is covered in two corresponding exercises. A common approach with PHA training is to alternate from upper body to lower body and back again. This is different from many other circuit-style approaches. This means that the exercise might follow this circuit shifting from upper body to lower body movements after doing a sufficient warmup: for example, 15 pushups, 15 calf raises, 15 pulldowns, 15 squats, and then repeating the cycle.

PHA Training improves resting cardiovascular changes and autonomic adaptation; in plain English, it constitutes a healthy, safe, and effective form of anaerobic exercise. The PHA method is similar to a circuit training method that keeps the exerciser moving from one exercise to another with minimal or no rest between exercises. In contrast, with PHA training, while the exerciser is moving from one exercise to another, the exercises alternate between upper and lower body exercises. For more details on PHA training and its effects, please see the article by Piras et al (2015) entitled "Peripheral heart action (PHA) training…" in the *European Journal of Applied Physiology*; and, view the youtube.com video entitled "SERGIO OLIVA'S FIRST BODYBUILDING ROUTINE-PHA SYSTEM! BLUEPRINT TO THE MYTH!!" at https://youtu.be/ogABJLruQu8 (April 15, 2019) and the youtube.com video entitled "How To Do Peripheral Heart Action (PHA) | Training System" at https://youtu.be/nyv9iSh16Z4 (May 10, 2020).

Part 6: Dark Comic Relief from Two Acclaimed American Restaurants

Two colorful American restaurants will be featured to provide some comic relief through twisted humor. Far more than these two culinary risqué (or high health risk) restaurants operate in the USA, but these two will serve our needs. To violate healthy practical vegan and vegan diets at least in degree, countless eating establishments can be frequented: for example, Outback Steakhouse, Texas Roadhouse, Burger King, McDonalds, Carl's Jr, Wendy's, Dairy Queen, Jack in the Box, Arby's, Cracker Barrel Old Country Store, Taco Bell, Chipotle Mexican Grill, Taco John's, Pizza Hut, Domino's Pizza, MOD Pizza, KFC, Buffalo Wild Wings, Golden Corral Buffet and Grill, Denny's, IHOP, Panda Express, Mongolian BBQ, Noodle Express, Sonic Drive-In, The Pita Pit, Olive Garden, Submarine, Dickie's Barbecue Pit, and numerous gas stations and diners selling hotdogs, corndogs, chilidogs, and so on. Ironically, hungry omnivores can eat unhealthily at numerous restaurants in the US, Canada, and elsewhere.

However, the two we choose to focus on offer succulent foods that tempt omnivorous and carnivorous gluttony openly and brazenly. The first one to be revealed is the "Big Texas Steak Ranch" in Amarillo, Texas; the second one is the "Heart Attack Grill" in Las Vegas, Nevada. As contrasted with the American Medical Association's suggestions for a healthy diet, these two restaurants flaunt the consumption of high cholesterol and greasy meat products in gross proportions - meals that put health at risk especially if such gluttonous meals were consumed routinely. The *Big Texan Steak Ranch* and the *Heart Attack Grill* epitomize two restaurants that cause pure vegans to have nightmares and practical vegans to laugh at the follies of the gourmands who dine there. Of course, there are many more than these two restaurants that serve excessive amounts, but we will look at these two due to their colorfulness.

The photographs of the restaurants and most of the information on these two restaurants can be seen in Wikipedia.com entries for "The Big Texan Steak Ranch" and the "Heart Attack Grill" (or "HAG") as well as through www.bigtexan.com/72-oz-steak and www.vegasfoodandfun.com/heart-attack-grill. After visiting both of these restaurants, Fiordo offers comments and impressions. The science author of this book, the cardiologist Dr. Noah

Chelliah, does not consecrate either restaurant. However, he would maintain that for someone in good health dining at such restaurants on rare occasion, the chance of that person dying or having any condition more than indigestion is slim. The healthy person who chooses to be a gorger on rare occasion at either of these restaurants will likely survive the tasty and enormous yet chancy meal. Neither Fiordo nor Chelliah has eaten at either of these enticingly ravenous restaurants. Nabal speaks his mind in this way: "These are my kind of restaurants. They are not restaurants for slackers and culinary cowards. Step aside so that I can show you how to eat huge portions free of charge!"

The *Big Texan Steak Ranch,* a steakhouse restaurant in Amarillo, Texas, has a large bull statue that advertises their "free" 72-ounce (2.04 kg or 4.5 pound) steak if a customer takes and wins the challenge of eating it and all that comes with it. The steak is free to anyone who, in one hour or less, can eat everything in the meal. The meal consists of the steak itself, a bread roll with butter, a baked potato, shrimp cocktail, and a salad. If a customer fails to satisfy the conditions of this eating challenge, they pay $72 for the effort. Those who have successfully consumed the 72-ounce steak challenge have their names recorded and posted at the restaurant. Since this challenge was instituted, around 68,000 people have attempted this challenge with 9,800 successfully devouring this gourmandizing feat: a rough calculation profit is $4,190,400. The challenge started in 1960. Those who take on the 72-ounce steak challenge are required to pay for the meal in advance. If they are successful, their money is refunded. The steak is cooked to the participant's preference, and the challenge takes place at a table for six on a raised platform in the middle of the main dining room. Nabal announces that this is his type of restaurant. He affirms: "I see no health risk at The Big Texas Steak Ranch even if I were to dine there daily!"

The website for The Big Texas Steak Ranch features a Hall of Fame section that displays all the names, dates, times, weight, height, and age of all the winners of the steak challenge. On 26 May 2014, the 125-pound competitive eater Molly Schuyler, set the record for the shortest time to finish the meal by devouring it in just 4 minutes 58 seconds. Schuyler returned on 19 April 2015 and beat her own record by 40 seconds. She had defeated competitors in the challenge, devouring two more meals in twenty minutes. The unofficial record for carnivorous wild animals was held by a 500-pound Siberian tiger who ate the steak in 90 seconds in 1999, until eclipsed by a lioness in 2012 clocking in at 80 seconds. Only a tiger and lion ate more than the most successful voracious and rapacious human omnivores. Nabal tried to outdo Schuyler and the lioness, but he got sick and had to drop out of the challenge.

After travelling through Amarillo seven times over a period of two years, out of curiosity on one trip through Amarillo, Fiordo decided to visit the Big Texan Steak Ranch. When he read the rules of the challenge and what was required to get the free gigantic meal, he controlled his carnivorous and gluttonous inclination to gamble on the 72-ounce steak challenge. He decided that the meal was truly monstrous in size and unhealthiness, returned to his automobile, and continued to travel westward and away from the restaurant. He was glad that he chose to exit rather than overindulge himself. Nabal called him unsavory names pertaining to his cowardice.

The *Heart Attack Grill* is an American hamburger restaurant in downtown Las Vegas, Nevada. It makes a point of serving food that is very high in fat, sugar, and cholesterol: food that, if eaten regularly, will increase the chance of a heart attack - therefore, the name. The restaurant was featured on several televised shows including "Extreme Pig-outs" on the Travel Channel.

The Heart Attack Grill establishment exploits a hospital theme in marketing its restaurant. Waitresses are cleverly called "nurses" and waiters are called "doctors" who take orders called "prescriptions" from the customers who are called "patients." Each patient wears a hospital gown and wrist band before ordering. Those who fail to finish their meal receive a paddling by one of the "nurses" and have the option to buy the paddle afterwards. Heart Attack Grill has deliberately courted controversy as a marketing strategy. The restaurant has been criticized for its flesh-revealing "breast-aurant" styled nurses who lure wanton diners in part with sex appeal.

The menu embodies the theme of items that violate health through meals that are high in artery clogging calories and fat: for example, "Single", "Double", "Triple", "Quadruple", all the way up to "Octuple Bypass" hamburgers. The gorger options include:

1) hamburgers ranging from 8 to 32 ounces (230 to 910 g) of beef up to about 8,000 calories (33,000 kJ)
2) all-you-can-eat "Flatliner Fries" (cooked in pure lard)
3) beer and tequila
4) "butterfat milkshakes"
5) soft drinks such as Mexican-bottled Coca-Cola made with cane sugar

Customers can violate more health standards and moral principles by ordering unfiltered cigarettes or candy cigarettes for children. To illustrate the extreme food choice of the Quadruple Bypass Burger, this burger has 9,982 calories (41,760 kJ) and consists of four half-pound beef patties, twenty strips of bacon, eight slices of American cheese, a whole tomato and half an onion served in a bun coated with lard! Nabal says: “What’s wrong with this? It sounds reasonable to me, and I know food. I know what’s good for me. I’m no fool! Not at all. I’m going to live to be a hundred and three. I play it safe for you and me because I’m no food fool!”

Customers who weigh in at over 350 pounds (160 kg) with a nurse waitress before eating get the meal free. One of the restaurant's promotions is a reward for customers who finish a Triple or Quadruple Bypass Burger. When customers finish the meal, they are rewarded by being placed on a wheelchair and wheeled out to their vehicle by their "personal nurse." Since the Heart Attack Grill markets its menu in a fatalistic way, some tragedies allegedly have followed the overindulgence of some customers. For example, on 11 February 2012, a customer suffered what was reported to be an apparent heart attack while eating a "Triple Bypass Burger" at the Las Vegas restaurant. Restaurant owner Jon Basso called 911, and the customer was taken to the hospital. Reportedly, patrons thought it was a stunt and started taking photos. Basso later said, "I actually felt horrible for the gentleman because the tourists were taking photos of him as if it were some type of stunt. Even with our own morbid sense of humor, we would never pull a stunt like that." And, in February of 2013, an unofficial spokesperson and daily patron, 52-year-old John Alleman, died of an apparent heart attack while waiting at a bus stop in front of the restaurant.

After traveling to Las Vegas several times over a period of three years, again out of curiosity, Fiordo decided to visit the Heart Attack Grill. When he entered the restaurant and got a sense of how the restaurant was run, he once again controlled his carnivorous and gluttonous inclination to gamble on the Quadruple Bypass Burger. Deciding that the meal was truly monstrous in its sick size, he returned to his automobile and drove to his motel to enjoy a swim in the motel's pool on a sunny day in the desert. He was glad that he chose the swim over the death-defying burger lunch. Nabal once again called him raunchy names pertaining to his gutlessness.

Part 7: Medical Science, Nutritional Science, and Practical Veganism: Broad Sweeps

Reflecting on nutrition, health, and disease (28 September 21), Dr. Chelliah has commented on diverse dietary and medical concerns. Experience in medical science, nutritional science, and practical veganism matters, and he asks us to *choose experience*—in particular, his extensive scientifically based medical experience. A brawny dietary, health, and weight control principle he stresses involves *calculating habitually a carbohydrate to fiber ratio of 5 to 1*. He maintains that *a 5 to 1 carb to fiber ratio is safe and commendable.* As the ratio increases to 10, 15, or 20 to 1, the healthiness of the food declines. We should check the carb to fiber ratio before consuming the food we plan to eat. We can do this by reading the labels closely. If tomato soup has a 5 to 1 carb to fiber ratio, then the carb to fiber ratio is safe and sound. If we want to eat white rice and the carb to fiber ratio is 20 to 1, then we might decide to choose another meal or add beans to the rice to increase the fiber content: for example, rice with black beans. Although adding beans or lentils to rice is helpful in aligning our choices with a 5 to 1 carb to fiber ratio, it is not ideal. It would be smarter to minimize or avoid high carbohydrate foods like rice, pasta, and potatoes. As far as nuts are concerned, Dr. Chelliah favors raw and unsalted nuts to roasted and salted nuts. A handful a day will suffice, whether we have the hand of a seven-foot professional basketball player or the hand of a five-foot gymnast! He ranks walnuts at the top since they have a carb to fiber ratio of 2 to 1 in addition to other heart healthy benefits. Other nuts that serve our healthy goals include the following in this order: almonds, Brazil nuts, pecans, pistachios (good also for sexual performance enhancement), cashews, and pine nuts. He stresses that cashews and pine nuts may be very tasty, but both are low in nutrients. Although not a nut, as a legume and not a tree nut like walnuts and almonds, *peanuts benefit us because of their arginine content which can curb our appetite and help us lose weight.*

An illustration of carbohydrate to fiber ratios might help elucidate various ratios. We use a vertical continuum to show health giving versus health taking ratios of carbs to fiber ratios. Starting with a nourishing 5 to 1/5:1 carb to fiber ratio, we will stop at a non-nourishing 25 to 1/25:1 carb to fiber ratio.

Scale of Carb to Fiber Ratios from Health Giving at the Top to Health Taking at the Bottom

5 to 1

10 to 1

15 to 1

20 to 1

25 to 1

30 to 1

n to 1

As for specific dietary spice supplement items, *Dr. Chelliah currently recommends that turmeric, garlic, ginger, saffron, and fenugreek be present daily* in our diet to ward off diverse diseases. While he cautions on the overuse of spices like black pepper, he ratifies its use with turmeric in the amount of a mere "pinch" because it boosts the power of turmeric. Nabal does not side with the advice to have but a pinch of black pepper. Nabal asserts: "A pinch of black pepper is good, a handful of black pepper best!" While a lot of black pepper is not healthy and a pinch of black pepper added to turmeric may counterintuitively be effective, Nabal has a right to his unscientific and unmedical opinion. While the power of the pinch may be counterintuitive, a "pinch" of black pepper added to turmeric can increase the power of turmeric by up to 2,000%! Turmeric can be added to foods like stews and soups; it improves conditions such as arthritis, inflammation, and dementia. Turmeric is comparatively inexpensive and can be purchased at big box stores.

Dr. Chelliah praises garlic on several grounds. It has anti-cancer potential and lowers cholesterol. He commends ginger as well because it reduces inflammation, improves digestion, lowers PMS, reduces migraines, and may lesson arthritic pain. Ginger decreases body weight and increases HDLs (the so-called "good cholesterol"). Daily consumption of ginger can also reduce menstrual blood loss by 50%! Saffron is sanctioned since it decreases memory loss and elevates mood. Fenugreek has benefits as well. Among its benefits are the lowering of blood sugar, the killing of cancer cells, and the improving of muscle strength. As for cardamom, it may improve stomach ailments and heart problems, and it may keep cancer from developing. Our caring physician also warns of spices that should be avoided: nutmeg which may produce hallucinations and tarragon which can be carcinogenic in large amounts. Since 500 people in the US die each day from cancer, to lower the risk of cancer and heart disease forming, Dr. Chelliah supports the consumption of a handful of nuts or seeds daily. Nabal is concerned because of the size of his hands. He complains in this way: "I'm only 5'4" but my hands are the size of Michael Jordan's. Does this mean I get to have only half a handful of nuts or seeds since my hands are twice as big as the hands of other people who are my height? If I get to have a full handful, I am finally getting a big fat break, or am I just getting gaslighted?"

Cloves and Ceylon cinnamon function as antioxidants. As a vasodilator, cocoa should be consumed without sugar or cow's milk. Non-sugars like erythritol or allulose and plant-based milks like oat or almond may be used with cocoa without significantly inhibiting its vasodilating powers. Cow's milk is high in cholesterol and oftentimes suffers from DDT contamination as a result of what cows eat. The oat or almond milk should be unsweetened so that erythritol or allulose can be used for sweetening. In its healthy form, we might amusingly refer to hot cocoa as "Hallmark cocoa" since Hallmark love stories and Christmas stories make extensive use of "hot cocoa" as a civilized and comforting drink. Of course, Dr. Chelliah esteems teas for their multiple benefits but especially certain ones for lowering blood pressure. For their health benefits, he ranks them from most to least powerful in this order: hibiscus, white, green, and black. He warns that hibiscus has the drawback of being highly acidic and can erode tooth enamel. Subsequently, he advises drinking hibiscus tea through a straw, drinking water after consuming it, or rinsing the mouth with water after drinking it.

Truth be told, Dr. Chelliah is quite sad about the Standard American Diet (SAD), for it falls short on an array of nutritional items. Overall, he advises us to *avoid extracted supplements for nutrition found in plants and go with actual plants instead.* He does not want us to rely on or substitute the product derivatives for the real product; hence, he does not want us to rely on or substitute juices and oils for the real things. With fruit juices and oils, he urges us to prefer oranges, apples, watermelons, etc. to juices flavored with these fruits and to prefer olives, avocadoes, etc. to olive and avocado oil*s*. Fruit juices tend to be high in sugar and are not in our best health interests. Many oils are useful in regulated amounts for cooking and for flavoring salads and other foods. A partial list of potentially gainful oils includes these: algae, olive, avocado, almond, walnut, hemp, sunflower seed, peanut, basil, grapeseed, safflower, and soybean. Dr. Chelliah puts

Thrive, a bit pricey culinary *algae oil*, above other oils because it is best for the heart. It appears to have the highest amount of monounsaturated fats and the lowest amount of saturated fats.

Of course, nicotine and fentanyl have some positive as well as negative effects. While nicotine causes blood vessels to constrict, it also results in lower incidents of Parkinson's disease. Recognize that nicotine occurs in tiny amounts in certain nightshade vegetables. While the US has declared war on fentanyl, legal fentanyl is on the World Health Organization's list of essential medicines when used in safe dosages as a pain killer. Nabal thinks in broad, groundless, oversimplified, and goofy categories. He sees countless benefits in nicotine and fentanyl and has them daily. He declares: "They have made me the magnificent specimen I am today."

Some readers will be disappointed in what Dr. Chelliah says the scientific research reveals about eggs and chicken. Eating eggs and chicken may lead to an increase in cancer. More than the regular consumption of beef or fish, regular consumption of chicken results in the greatest increase in weight gain. But, for fish lovers, is fish healthy? Nabal says: "Absolutely, yes." But, then, what does Nabal know? Sadly, fish are high in PCB toxins and dioxins as well as cholesterol. Subsequently, fish should be avoided, especially by women during pregnancy. Fish has a cholesterol rating of 70, shrimp of 125, and oysters 206. Fish face contamination from toxins and dioxins through *biological amplification*: toxins and dioxins accumulate up the food chain—thus, bigger fish eat smaller fish. Toxins accumulating in the smaller fish get transferred to the larger fish. Contamination in the seas is worldwide, even in seas to the far north. Therefore, should we turn to farmed fish? Nabal would, but, counterintuitively, farmed fish has more contaminates than wild caught fish. Toxins and dioxins increase in fish through these causal steps thanks to biological amplification. Like a snowball rolling downhill and gathering mass, small fish may be collecting toxins from the water. Then, small fish are eaten by medium size fish. Medium size fish are eaten by large fish. By the time we eat salmon and tuna, we have toxins galore from fish feeding on one another from the smaller to the larger on the fish food chain. In short, salmon and tuna, two popular menu items can be filled with toxins.

Nabal cries out: "If there is no solution, then go for the gusto. Eat whatever you like to your heart's content!" Dr. Chelliah reminds us that there is a solution to fish contaminates. *The main health benefit cited for consuming fish is the Omega-3.* Get Omega-3 not from fish oils but from algae—which is plant based. Let us emphasize that the fish oil business is a billion-dollar industry. Should we follow the money to determine whether the source is being honest or not? In this case, a billion-dollar industry may support those who claim fish oil for Omega-3 is healthy, whether or not the science supports it. So, Omega-3 from plant-based sources can minimize the poisoning from biological amplification. For example, Freshfield Vegan Omega-3 premium algae oil is plant-based, mercury free, and affordable. In comparison with conventionally marketed Omega-3 from fish oil, being mercury free, it lacks the biological amplification risks of fish oil. Nature Made Algae Omega-3 is an economical option for plant-based Omega-3; its ingredients are comparable to the Freshfield Vegan Omega-3. Both are worthwhile alternatives to Omega-3 from fish oil.

Two studies might shed light on the issues raised by using fish oils to improve heart conditions—the DART-I (Burr et al, 1989) and the DART-II (Burr et al.2003) studies. *DART* is shorthand for "diet and reinfarction trial." Many physicians and patients refer routinely to the DART-I study but fail to refer also to the DART-II study. The DART-I study reported in 1989 found positive effects from heart patients with angina consuming fish. The Omega-3 fish oil industry was jubilant. The study supported their industry. Sadly though, in 2003 the DART-II study was published by the same team of researchers that published DART-I; they did not find data supporting the beneficial findings on fish oil consumed by patients with angina reported in the DART-I study. Angina sufferers and their cardiologists had to face in the DART-II study findings that failed to show benefits from the fish oil dietary advice given to those living with angina. In fact, it showed that *fish oil caused harm.* Nabal still sides with the DART-I study, for the results were warm and fuzzy for his fish-eating preferences for acquiring Omega-3. The truth can be such a scourge to Nabal and his associates who shout out in unison: "Don't confuse me with the facts! Correlation and causation are the same!"

Dr. Chelliah calls much attention to diet and cancer. He reminds us that cholesterol promotes cancer, and cancer thrives on cholesterol. Subsequently, he cautions against the consumption of lamb, beef, eggs, cheese, and other animal sources for meals and animal products. He emphasizes that cooked meats develop heterocyclic amines that may eventuate in cancer. Barbequing meat, more than conventionally cooking meat, increases the danger of cancer from heterocyclic amines. To counter cancer, Dr. Chelliah prescribes consuming cruciferous vegetables and lignans found in many plants since both kill cancer cells. Cruciferous vegetables encompass kale, cauliflower, cabbage, garden cress, bok choy, broccoli, Brussels sprouts, mustard plant, and similar leafy green vegetables. Lignans can be found in flaxseeds, sesame seeds, cereals (such as wheat and oats), soybeans, tofu, cruciferous vegetables (like broccoli and cabbage), and some fruits (prominently apricots and strawberries).

Cruciferous vegetables kill cancer cells in the breast, stomach, prostate, colon, kidney, and bladder. Ingesting whole grains, nuts, seeds, berries, apples, pears, persimmons, dark green and dark colored vegetables, beans, kale, spinach, and broccoli contribute to the death of cancer cells. For maximum nutrition, there is a chemical effect that enhances the natural curative powers of such cruciferous vegetables as broccoli, cabbage (red or purple over green), cauliflower, Brussels sprouts, bok choy, kale, collards, mustard greens, and watercress. For instance, to produce this beneficial chemical effect with broccoli, Dr. Chelliah advises that broccoli, fresh or frozen, be chopped before eating it because sulforaphane is a compound produced by damage to the plant—as from chopping. The chopping allows two compounds to mix and react for our nutritional enhancement. Chiefly for women, breast cancer can be countered by eating cooked mushrooms, *not* raw mushrooms. Agaritine is in raw mushrooms and is a toxin. Because heat kills the potentially toxic agaritine, Dr. Chelliah prescribes cooking the mushrooms. There is no need for fancy mushrooms like shitake or portobello. Plain white mushrooms will suffice for the protection of breasts from cancer. Eating cooked white mushrooms daily makes everyday a breast cancer awareness and prevention day.

Sweet potatoes, too, inhibit cancer growth. In 1931, a unique protein was discovered in sweet potatoes. Eighty percent of the protein in sweet potatoes constitutes a protease inhibitor which yields anticancer effects. The protease enzyme promotes cancer growth in these body parts: tongue, cervix, colon, and rectum. Enjoying sweet potatoes, thanks to their protease inhibition, can keep cancers from growing in us. Therefore, a sweet potato a day may very well keep the cancer away—along with the doctor! Nabal rants against sweet potatoes: "They are useless. Give me a good old Russet potato mashed up with the skins still on them—not an ugly old sweet potato. Don't tweet sweet potatoes to me. Tweak them instead with Russet spuds and their skins with lots of butter, salt, and black pepper."

When we check Wikipedia.com under "heterocyclic amines," we find a treasure of information on this topic. The high temperature cooking of meat and smoking of plant matter like tobacco create 27 heterocyclic amines. Cooking meat at high temperatures (especially charring meat) forms some cancer-causing heterocyclic amines. Some heterocyclic amines that are found in cooked (and markedly burned) meat are carcinogens. Heterocyclic amines occur in meat at high cooking temperatures when cooking muscle meats like beef, pork, fowl, and fish. The United States National Cancer Institute's Division of Cancer Epidemiology and Genetics discovered a link between stomach cancer and the consumption of cooked meat. Other studies have uncovered links between high intakes of well-done, fried, or barbecued meats and colorectal, pancreatic, and breast cancer. Interestingly, people eating medium-well or well-done beef were more than three times as likely to suffer stomach cancer as those who were eating rare or medium-rare beef. Intriguingly, milk, eggs, tofu, and organ meats like liver have little or no heterocyclic amines naturally or when cooked although they have a number of other health problems associated with them. Strikingly, *olive oil, lemon juice, and garlic marinade cuts heterocyclic amine levels in chicken by as much as 90 percent while marinating a beef steak in beer or red wine for six hours cuts heterocyclic amine levels by as much as 90 percent compared with unmarinated steak.* Alluringly, a bit of preparation through marinade and marination of meat prior to cooking it goes a long way in reducing the provocation of cancer through this one mechanism. But other problems with animal products remain, such as stimulation of the secretion of IGF-1 which promotes cancer. This particular method of carcinogenesis is unaffected by cooking or marinating meat. Nabal wants to know if it is ok to drink the wine and have a beef steak marinate in his tummy. Sorry, Nabal, Dr. Chelliah says the steak has to be marinated before cooking it. Good try though!

Additional information is provided by Dr. Chelliah for preventing cancer. Although he knows that his medical suggestions to prevent cancer may be disturbing to many, his advice for optimal benefits include:

1. Walk four miles per day. *The role of exercise in supporting health cannot be overestimated.* Four miles of walking per day equals a sizable amount of exercise.
2. Avoid all animal products.
3. Consume three tablespoons of flaxseed daily. Flaxseed reduces the incidence of prostate cancer, breast cancer, and colon cancer. It also *lowers cholesterol levels.*

4. Eat eight servings (four cups) of dark greens daily.

5. Eat four servings (two cups) of fully cooked beans or lentils daily.

6. Consume turmeric, ginger, and garlic daily.

7. Eat sweet potatoes, purple cabbage, cooked mushrooms, and cauliflower regularly.

8. Avoid high fat items and sugar. In particular, *avoid fructose* from all sources except from fruit such as berries.

9. Sleep seven (not 4 or 9) hours at night in as dark a room as possible. Use a clothespin, if necessary, to connect drapes in order to seal out the light from partially opened drapes. Significantly, not getting enough sleep leads to cancer, heart disease, high blood pressure, stroke, and diabetes. *Melatonin helps destroy cancer.* Nuts and seeds contain melatonin. Nature's sources of healthy minerals, vitamins, etc. are best. *Supplements are not as beneficial to our health as that which we absorb from natural foods.* The pineal gland produces melatonin *only when there is no light shining into the eyes.* It was found that because blind women have no light hitting the retina, they have very high levels of melatonin and hence fewer breast cancers and possibly other cancers too.

10. Eat berries daily. Eat them plain with no whipped cream. Strawberries can be used to treat early esophageal cancer. *Wash berries thoroughly in salt water and then rinse them. At least rinse berries in water to rid them of any harmful chemicals that were sprayed on them.* Eat all the berries you desire. Nabal finally sees a silver lining, for he loves berries. Being kind, Dr. Chelliah praises Nabal and tells him to go for it. In berries may be the salvation of Monsieur Nabal!

Miscellaneous health items will now be discussed. Dr. Chelliah supports eating white potatoes, but he discourages us from eating the potato peal because there is a toxin in the peal. Purple potatoes (the flesh is purple in color) are much more nutritious than white potatoes. Eating the skin of sweet potatoes is acceptable. Consuming beans daily gives us protein, carbs, fiber, and longevity. As Dr. Chelliah repeats, *a diet free of beans is suicidal.* Additionally, a life without beans may result in obesity, a decline in health, and a death sentence. *People who love and eat beans regularly are the longest living people in the world!* A sad point must now be made. The *beans should not have pork, sugar, honey, or molasses in them.* The beans can be sweetened with erythritol or allulose. Erythritol is a bit costly and sweet (but not as sweet as sugar), yet it is an antioxidant and has no calories. *Dr. Chelliah playfully suggests that only people who do not have any skills in cooking might add a sweetening agent or meat to cooked beans.* The slight bitterness or off taste of beans can be easily overcome by using onions along with a judicious use of spices (such as turmeric, garlic, ginger, and cumin). *With canned beans, choose the low salt beans and do not drain the liquid.* Cook or ferment the beans completely to kill the toxic lectins. To get a sweet taste, Dr. Chelliah suggests cooking the beans with onions. Honey *contains fructose which* promotes cancer and *weight gain.* Regretfully, in the US, fructose is a leading cause of diabetes, heart disease, and possibly cancer.

We must remember to eat, especially the beans we savor, so that we keep eating beans. Again, we should always eat as close to nature as possible. For example, ideally, we should eat cooked soybeans, when possible. If cooked soybeans are not available, we may consume tofu instead. If both are not available, we may consume soy isolate in small quantities because it is highly processed. Fiber is lacking in the Standard American Diet and that is truly SAD. Only three to four percent of us in the US consume enough fiber. *Fiber in beans will lower cholesterol levels. Cholesterol rates increase from Japan to the US.* In Japan, people indulge healthfully in a lot of edamame (high in fiber) and tofu—*with the edamame being closer to nature than tofu.* When we dine on beans and lentils, we eat less and may lose weight.

Fiber is a carbohydrate that human beings are not able to digest because we do not have the enzymes to digest fiber. Per contra, the billions of good bacteria in the colon have the enzyme to break down the fiber. When the bacteria digest these fibers, they release several chemicals which are highly beneficial to human beings. We will focus on just two of these chemicals: 1) propionates which increase satiety and hence make us eat less at the next meal! 2) butyrates which are critical for maintaining the integrity of the intestinal lining among other good effects. Of course, one of the byproducts of this digestion of fiber in the colon is the release of CO_2 which is the gas people experience.

With bean consumers, for a while, some humans will produce more carbon dioxide (that is, gas) than others. The gas is produced in the colon. Fortunately, *we profit biomedically from the lentil-bean effect, for we eat less.* To reduce gas, Dr. Chelliah recommends sprouting the beans or adding baking soda to the beans or lentils when soaking them. Importantly, seek a measure of the carb to fiber ratio, recalling that pasta may have a bad ratio rating of around 16 to 1 while low salt garbanzo beans may have a good rating of around 5 to 1. Also, as a memorandum, we can also combine foods to increase our carb to fiber ratio—thus, rice combined with beans (or soybeans) and wholewheat pasta combined with broccoli or spinach. In practice, ideally, each item on your plate should satisfy the 5:1 Carb to Fiber rule. But *a compromise would be that at least the whole meal should satisfy the 5:1 Carb to Fiber rule.*

To underscore the sense and sensibility in Dr. Chelliah's practical vegan approach to eating, he acts pragmatically to deal with patients with serious illnesses at advanced stages. *If a heart patient needs stents to open an artery, he prescribes stents.* If a cancer patient needs chemotherapy to control the growth of cancer, he prescribes chemotherapy. Nonetheless, *once the patient has been medically treated through conventional surgery* or chemotherapy (and sometimes while the patient is preparing for the conventional medical treatment of the disease), *Dr. Chelliah enjoins the patient to follow his practical vegan diet, if not his vegan diet*, to avert the recurrence of the disease or worsening of the disease.

Part 8: Vegan Imitations of Animal-Based Foods

Many commercial substitutes for meat have emerged in the last few years. Some are thoroughly tasty meat imitations created to replace hamburger, chicken, seafood, milk, cheese, eggs, and butter. We prudently support eating substitutes for animals and animal products *in small quantities* whether the substitutes are tasty or not to those consuming them. Here are some animal-based imitations we have tried and would recommend to you for tasting and possible consumption but with the following caveats: 1) Watch out for high sodium content; 2) Watch out for high fat content, particularly saturated fat and particularly coconut oil.

Please consider these vegan alternatives:

1) Loma Linda products (especially, their kosher franks, chicken nuggets, and scallops).
2) Beyond Meat products (especially, their Cookout Classic plant-based burger patties).
3) Good Catch (especially, their plant made fish free tuna).
4) Tofurky Company's all vegan, plant-based ham style roast with or without amber ale glaze.
5) Califia Farms non-dairy milk: oat, almond, and others.
6) Violife Cheese (especially, their vegan mozzarella imitation cheese).
7) Orgran (especially, vegan easy egg for cooking).
8) Upfield butter (especially, I can't believe it's not butter—light).
9) Eat Just products (especially, Just Egg).
10) Sweet Earth's Mongolian Beefless.
11) Higher Taste's Happy Hoagie (especially their Kahuna—a tasty vegan teriyaki with rainbow slaw) out of Cornelius, OR (www.thehighertaste.com).
12) Rebbl (at rebbl.co) vegan non-dairy milk: especially their dark chocolate and their turmeric. The ingredients Rebbl uses are mostly nourishing and include the following for chocolate and turmeric. The chocolate non-dairy milk includes plant protein, maca, reishi, ashwagandha, and MCTs. *(Dr. Chelliah recommends against MCT).* The turmeric milk *(non-dairy)* includes turmeric, ginger, and MCTs. Both vegan milk products boast that they are USDA Organic and have no soy, dairy, GMO, gluten, cane

sugar, or artificial colors. Both products come packaged in a 100 percent recycled plastic bottle. The Rebbl offerings constitute premium quality non-dairy milk products.

We judiciously endorse and consume the above substitutes that are found at "vegan" restaurants, and we are grateful for the advances in taste that several companies have achieved. However, we do not limit ourselves to the vegan substitutes noted above for animal products. We also prudently approve vegan products that are natural and not meat imitations: for example:

1) Tasty Bite (like their organic Indian Madras lentils).
2) Seeds of Change (like organic quinoa and brown rice with garlic). For those who like Indian Madras lentils from Tasty Bite but find the flavor too strong, we suggest they mix the Madras lentils with quinoa or brown rice to dissipate the intense flavor of the Madras lentils.
3) Safely and deliciously prepared cereals (like Barbara's Shredded Wheat by Three Sisters Cereal from Lakeville, Minnesota—a cereal that has 7 grams of dietary fiber, zero cholesterol, zero saturated fat, zero sodium, and zero sugar).
4) Munch Rite Brown Rice Cakes with 7 Super Seeds, a product of Bulgaria distributed by Golden Beach, Inc. (www.goldenbeachinc.net), for a snack or meal. Rice cakes can be eaten plain or with peanut butter, tahini butter, almond butter, etc. with sugar free jam or jelly.
5) Mulgipuder, a dish from Estonia, is a dish made with potatoes, groats, and meat (see mulgipuder in Wikipedia under *Estonia*). The amount of meat can vary.
6) Smoothies with diverse healthy vegan ingredients: for example, a blend of almond milk, bananas, crushed ice, cocoa, and erythritol. Orgain Organic Protein and Probiotics makes a creamy chocolate fudge flavor protein powder that is highly suitable for blending a smoothie. The protein powder by Orgain is plant-based, dairy free, gluten free, lactose free, non-GMO, USDA organic, and certified vegan. Orgain sweetens their vegan product with organic erythritol and organic stevia. ***Dr. Chelliah cautions us against the consumption of stevia.*** Fiordo interviewed a publisher of a health magazine and a lifelong health enthusiast. He told Fiordo that to control his appetite and obtain sufficient nutrition, he creates daily a smoothie mix of the type just described and consumes a quart of it over the day. He consumes numerous healthy variations of a chocolate smoothie in addition to the one mentioned. Fiordo also interviewed a professional Muay Thai fighter and trainer who favored a blended quart of this smoothie: blueberries, crushed ice, alkaline water, erythritol, mixed nuts (salt free), and a fruit flavored vegan protein powder. He also liked many healthy variations with other berries for similar mixes.

7) "Chelliah Chips," a tasty and healthy snack created by Dr. Chelliah to satisfy hunger, control weight gain, and improve digestion.

For practical vegan dieters using a plant to meat ratio of at least 80:20, we support watchfully those trying to decrease, but not eliminate entirely, their quantity of meat by consuming meat alternatives: for example, soups with small amounts of chicken, such as chicken noodle soup. We also favor the consumption of foods like beef fried rice from Chinese restaurants, sushi with salmon from Japanese restaurants, or sushi-like kim bap from Korean restaurants. Chinese restaurants also offer vegetable fried rice and Japanese restaurants offer sushi without meats. Currently, some popular fast-food restaurants like Burger King now offer plant-based burgers: namely, the "Impossible Burger." Of course, abundant vegan options can be found in Italian, Chinese, Japanese, Thai, and Mexican restaurants as well.

Part 9: International Vegan Cuisine and Other Eating Options

a Hints for tasty international cuisine options

Let us remember that we are in the revered presence of our inspired epigram that applies to consuming food: *Good food must first get past our lips to do us any good!* While we support the consumption of delicious cuisine vegan or practical vegan meals that imitate animal-based foods, we embrace savory traditional vegan meals that do not imitate animal-based foods. Many tasty traditional vegan meals from around the world can be consumed at diverse restaurants, ethnic and conventional, in the US: that is, meals that get past our lips! A selection of traditional cuisine vegan meals that can use vegetarian ingredients and vegan ingredients include: potato pirogis, miso soup, meat free baked beans, salsa for dipping corn ships, spaghetti (and numerous other types of pasta like rotini and orecchiette) in marinara sauce, spaghetti with olive oil and garlic (sometimes with red pepper, parsley, parmesan cheese, and pecorino cheese added), vegetable minestrone soup, beet soup, botwinka soup from Poland, khorta from Greece, preboggion from Italy, rice and fufu from Nigeria, saag from India, Bhindi masala from India, sushi without meats, Daal makhana from India, Biryani with vegetables, Roti (wholewheat vegan friendly bread) from India, Akara from Togo, Lobia curry (a black-eyed peas dish from India), vegetable fried rice, twice baked potatoes, sweet potatoes, warm tofu (soybean curd) with garlic sauce, tempeh made from fermented soybeans (for example, tempeh goreng, fried tempeh, grilled tempeh, and tempeh sandwich), Red red from Ghana, French fries (especially in an air fryer), Norwegian potato lefse made with vegan ingredients (of flour, non-dairy milk, and non-dairy butter), Norwegian polse med lompe (sausage with lompe, a type of potato lefse) made with vegan ingredients (vegan sausage or vegan hotdogs replacing those made of pork or beef), quinoa salad (including such optional additions as artichoke hearts, beets, olives (black or green), spinach, avocados, potatoes, sweet potatoes, dry edamame, sunflower seeds, pistachios, various nuts, raisins, and sandwiches made from peanut butter, cashew butter, pecan butter, tahini butter, or almond butter with low or no sugar jam or jelly.

A selection of traditional cuisine practical vegan (that is, semi-vegetarian) meals will be mentioned next. The meals may blend plant and animal products with a PM ratio of 70:30 and sometimes with a PM ratio as high as 99:1! Some options that are available include: cheese ravioli, frittata, saag gosht (spinach and goat), beef fried rice, shrimp fried rice, and chicken fried rice. Quinoa can be easily mixed with rice (especially brown rice) with diverse beans added for fiber—for example, lima, baked, black, or garbanzo beans. In place of beans, lentils can be added to the rice or to the rice and quinoa.

b Hints for tasty vegan options

Assuming with some hesitation that most people are not chefs, several simple cuisine meals follow. Rice, especially brown or basmati, can be used with or without quinoa, lentils, or roasted garlic to create meals. Vegetables can be added to the mix: cooked carrots, spinach, broccoli, kale, onions, and other flavorful choices. For a quick meal, we can microwave for 60 seconds a product like Organic TruRoots' roasted garlic quinoa and brown rice with Tasty Bite's organic Indian madras lentils. The madras lentils mix easily and palatably with the quinoa and rice. A fine substitution can occur with stuffed cabbage rolls. Instead of stuffing the cabbage leaves with ground meat, they can be stuffed with ground plant burgers like those marketed by Impossible Burger or Beyond Beef. "Meatless" meatballs can be blended with marinara sauce with different pastas as a tasty vegan option. Similarly, tater tots and veggie tots (using broccoli, cauliflower, and other more nutritionally dense vegetables) can be cooked with plant burgers instead of ground meat. The taste of these dishes for many may be close enough to meat that a substitution of the plant-based meatless burger for meat may be delicious for many people.

c Hints for tasty practical vegan options

Most of the meals covered under international cuisine and veganism can be toothsome when served with meat. To remain sensible in preparing practical vegan meals, cooks should do their best to stay in a 10 to 20 percent range for quantity of meat: thus, a plant to meat ratio of 80 to 20 or 90 to 10. To be specific, vegetable fried rice can be transformed into an 80:20 PM ratio by adding 20 percent beef to make it beef fried rice. Staying with easily prepared meals, the blend of quinoa and rice with madras lentils noted above can be eaten on a strictly vegan basis or on a practical vegan basis. If we keep the practical vegan plant to meat ratio high (at 80:20 or 90:10), we can add lamb, goat, bison, deer, beef, or other meats to the vegan options described above: for example, tater tots, veggie tots, cabbage rolls, and a blend of roasted garlic quinoa and brown rice.

To do our best to trick our taste buds, some clever and even sly combinations are possible as well. We might have a PM (plant to meat) double cheeseburger by having one patty made from ground beef and one patty made from Beyond Beef. Instead of dairy cheese, we can use vegan cheese. To the combined PM cheeseburger might be added onions, tomatoes, mushrooms, and spinach—all of which are cooked. Also, the PM cheeseburger can have two unequal patties with the plant patty larger than the meat patty. For those who like chicken nuggets, a similar strategy can be deployed: that is, mix the chicken nuggets with the vegan imitation of chicken nuggets for a PM blend of nuggets at an 80:20 ratio! The taste differences may be for many indistinguishable. With imitation hotdogs, to trick our tastebuds, we might mix actual beef wieners with imitation vegan hotdogs in something like Bush's vegetarian beans—again, with an 80:20 PM ratio.

Part 10: Luscious Vegan and Practical Vegan Recipes from Dr. Noah Chelliah

From his vegan kitchen, Dr. Chelliah offers the following recipes for delicious and healthy food to our readers. Many of his patients and friends have dined with him in his kitchen, judged these dishes to be superb, and ingested them regularly with gustative pleasure. After trying these tasty, wholesome, and spicey dishes, we are confident that you will give them rave reviews and enjoy them often.

BABY LIMA BEANS

Ingredients:

1. Onion 1 Medium
2. Mustard seeds (whole) ½ tsp.
3. Baby lima beans (frozen) 1 packet
4. Salt to suit your taste
5. Vegetable oil as needed
6. Unsweetened coconut flakes 1 tablespoon

Optional:

1. For those who like it spicy, you may add a small amount of red chili pepper powder and sauté it for one minute just before you add the baby lima beans.
2. May also add a pinch of turmeric at the same time with or without red chili pepper powder.

Method of Preparation:

Heat 2 tablespoons of oil in a dry saucepan. Add mustard seeds to the oil and immediately cover the saucepan! As soon as the mustard seeds start to pop, turn off the heat so that the mustard seeds do not burn!! Add the chopped onion and sauté it for a few minutes. Add the frozen baby lima beans and cook until it is barely done. Cook only for a few minutes. DO NOT OVERCOOK! Add salt for taste. Just before serving, sprinkle the coconut flakes on the cooked baby lima beans.

Serves 4-6

Serve as a vegetable side dish.

SPINACH STIR FRY

Ingredients:

1. Spinach - 1 lb.
2. Oil - 1 tsp.
3. Bengal Gram dhal / Channa dhal - 3/4 cup (smaller version of garbanzo beans—skinless split)
4. Urad dhal (skinless split black gram) - 1 tsp.
5. Red dry serrano peppers - 2 (may use less or skip completely)
6. Garlic - 4 pods
7. Ginger 1" finely chopped
8. Mustard seeds - 1/2 tsp.
9. Red onion - 1 large (finely chopped)
10. Asafoetida / Hing - a pinch
11. Turmeric powder - 1 tsp.
12. Cumin seeds - 1/2 tsp.
13. Grated coconut unsweetened - 2 tablespoons
14. Salt to suit your taste

Method of Preparation:

Soak the Bengal Gram dhal for 15 mins and cook it. Then, keep it aside.

Heat the pan with oil.

Add mustard seeds and let them pop. As soon as the popping begins, turn off the heat. *Burnt mustard seeds taste terrible!*

Add cumin seeds and let it cook for 30 seconds.

Add Urad dhal, red chilies, and Asafoetida. Fry to a brown color.

Add crushed garlic, ginger, and onion. Fry until transparent.

Add spinach, salt and turmeric powder and let it cook.

Add cooked Bengal gram dhal. Mix well. Cook for 1 min.

Add grated coconut and mix well.

Serves 6

Indian Lentils (Toor Dhal)

Ingredients:

1. Toor Dhal (split pigeon peas) 1 cup
2. Onion (chopped) 1
3. Green cayenne pepper (optional) 1
4. Cumin seeds ½ tsp.
5. Mustard seeds ½ tsp.
6. Garlic 2 cloves
7. Oil 2 tablespoons
8. Curry Leaves 8 leaves
9. Cilantro leaves 12-16 leaves
10. Salt for taste

Method of Preparation:

Rinse the Toor dhal in water at least twice; then boil them in 4 cups of water (or more as needed) until they are fully cooked and "mushy." Then, keep them aside.

Place 1 tablespoon of oil in a frying pan and heat it to medium heat. Add the chopped cloves of garlic and roast them for a minute or two. Then add the chopped onion (and chopped cayenne pepper if you use it) and sauté until the onion is beginning to turn brown. Now add the Toor dhal and cumin seeds and roast the whole mixture for a minute or two. Add the chopped cilantro leaves and cook for 1 minute. Remove the mixture from the heat.

Heat the remaining oil in a fresh frying pan on high heat. When the oil is hot, add the mustard seeds and immediately place a lid on the frying pan. Listen to the mustard seeds pop. As soon as the mustard seeds start popping lower the heat, so the mustard seeds do not burn. Then add the curry leaves and roast them on medium heat for a minute or two. *Don't let them burn!* Now add this oil with mustard seeds to the Toor Dhal mixture previously made. Stir well, add salt, and serve.

<u>Serves 4</u>

This can be eaten with white rice, multigrain bread, or by itself.

Beans

This recipe can be used for almost any type of dry beans. Dry beans include:

Chickpeas (garbanzo beans)

Blackeye beans

Kidney beans

Moong (Mung) beans

To prepare any of these dry beans, you must soak them in cold water over night. *Wash them before soaking*! Blackeye beans and kidney beans need to be soaked only for a few hours.

Ingredients:

1. Anyone of the above beans: 3 cups
2. Coconut grated: 1 cup
3. Mustard seeds: 2 tablespoons
4. Onion: 1 big (chopped)
5. Green chili (cayenne or serrano Peppers): 4 (cut into small pieces) (Optional)
6. Curry leaves: (2 sprigs) You may substitute this with cilantro leaves (chopped)
7. Oil: 2 tablespoons
8. Asafoetida: 1/4 tsp. (*if you have it*)
9. Water: As needed

Method of Preparation:

Cook the beans with water in a 1:2 ratio of beans to water. Beans should be cooked completely but don't overcook the beans. Drain excess water. Add a little salt towards the end of cooking if you desire. Heat oil in a skillet. Add mustard seeds and allow them to splutter (pop). Place a cover to prevent the splatter of oil. Add onion, chili peppers (*you may skip these if you prefer*), and curry leaves. Then sauté them until the onion is translucent. Add the beans and sauté the mix for 5 minutes so that the onion and green chili peppers get mixed well with the beans. Sprinkle powdered salt if needed and sauté the mix well. Now add the grated coconut and sauté it once again for a minute. Then, remove it from the flame.

Serve 6-10

PURPLE CABBAGE

Ingredients:

1 Medium sized cabbage finely chopped lengthwise

2 tsp. of Black Gram (split and skinless)

2 tsp. of Mung beans (split and skinless)

1 tsp. of ginger-garlic paste*

1 tsp. of whole mustard seeds

1 Large onion chopped

1/4 tsp. of citric acid (crystals)

1 tsp. of Lemon juice

1 tablespoon of oil

10-12 dried red long chili peppers (cayenne)

Salt for taste

2 Tablespoons of unsweetened coconut flakes

*You can use ½ tsp. of finely chopped fresh ginger and ½ tsp. of finely chopped fresh garlic instead.

Method of Preparation:

In a large saucepan, heat the oil and once hot, add the mustard seeds and cover it with a lid. As soon as the mustard seeds start "popping" turn off the heat. Once the mustard seeds stop popping, add the dried chili peppers and roast them for 1 minute.

Now add the ginger-garlic paste and sauté for 1 minute. Then add the chopped onions, black gram and Mung beans and roast them for 7-10 minutes or until onions are clear. Now add the cabbage and cook it until it is just barely done.

Salt for taste.

Just before you serve it, add the lemon juice. Mix it well. Taste it. You may need to add more salt to balance the sourness of lemon juice. Sprinkle coconut flakes and serve.

Serves 6

TOMATO SOUP (The Quick Version)

Ingredients:

1 Tablespoon of "The Chelliah Health Mixture" of spices. [Please see the next section for the recipe for The Chelliah Mixture.]

1 Bottle of Wal-Mart "Great Value" Tomato Basil Garlic Pasta Sauce

1 Bottle of Wal-Mart "Great Value" Low Salt (Low Sodium) Vegetable Juice

5 Drops of "Supreme Tradition" "Chicken Wing Sauce" sold at a "Dollar" store for $1

Method of Preparation:

Heat the saucepan on medium heat and add 1 tsp. of oil. Add 1 tablespoon of "The Chelliah Health Mixture" of spices and cook for 30 seconds ("blooming the spices"). Add the Tomato Basil Garlic Pasta sauce and the Low Salt Vegetable Juice. Then, bring it to simmer. Now add the 5 drops of chicken wing sauce and, if needed, add salt for taste.

<u>Serves 6-10</u>

May freeze left over soup.

Stir Fried Potatoes Indian Style

Ingredients:

Potatoes (red or white) 2 lbs.

Olive or canola oil 2 tablespoons

Whole mustard seeds (color does not matter) 1 tsp.

Red cayenne powder 1 tsp. (less if you like or skip it)

Turmeric 1 pinch

Salt 1 tsp. (or for taste)

Ingredients for a variation: (See below)

Cumin seeds ½ tsp.

Cumin Powder 1 tsp.

Cilantro Leaves 10-15 leaves

Method of Preparation:

Boil the potatoes just like you would for potato salad: that is, fully cooked but still firm not mushy. Peel the potatoes either before or after cooking. Drain the hot water and cover it in cold water until the potatoes are cold. Now cut the potatoes into small cubes and set aside.

Then heat a large skillet or saucepan with the oil. When the oil is hot (not smoking), throw in the mustard seeds and place a cover over the skillet. Listen very carefully. When the mustard seeds start popping vigorously, turn the heat off. Again, burned mustard seeds taste terrible!! When the mustard seeds have finished popping, add the potatoes, and turn up the heat. Now add the turmeric and cayenne powder. Roast the potatoes for 2-3 minutes, stirring them constantly.

Serves 4-7

Comments: For a variation on the above, you may add the cumin seeds. If you decide to use them, add the cumin seeds to the oil as soon as the mustard seeds have finished popping and roast these for 1 minute before proceeding. You may also add cumin powder. If you do, add it at the same time you add turmeric and cayenne powder.

For additional variation you may sprinkle the potatoes with coarsely chopped fresh cilantro leaves one minute before you remove the potatoes from the heat. Always add the salt and/or cilantro toward the end of cooking and just before you remove the potatoes from the heat.

Always peel the potatoes and discard the peels.

Coconut Basmati Rice

Ingredients:

1. Basmati rice 3 cups
2. Coconut milk (unsweetened) ¼ of a can
3. Cardamom 6 pods
4. Turmeric powder 1 pinch
5. Clarified butter ½ tsp.
6. Cloves 5
7. Cinnamon ½" Stick
8. Salt for Taste

Keep aside the following until called for:

9. Onion 1 large
10. Raw Cashew nuts 1/8 lb.
11. Clarified butter 2 tablespoons

Method of Preparation:

Add all of the ingredients numbered 1 through 8 into a rice cooker. Now add enough water to bring the level to mark 3 in the rice cooking pot (approximately 1 ½ cups of water to each cup of rice).

Remember that you can always add more water to undercooked rice, but you can never take away water from the overcooked rice!!

Plug the rice cooker into the power outlet and turn the switch to "cook" and not to "warm." Add 2 tablespoons of clarified butter in a saucepan. Slice the large onion lengthwise into long strips. Sauté the sliced onion strips in clarified butter at medium heat until the onion strips become crisp. Remove the onion strips and keep aside. Now, into the same saucepan with clarified butter, add the raw cashew nuts, and fry at very low heat. Do not burn the cashew nuts.

The rice cooker will automatically switch from the "Cook" position to the "Warm" position. Do not turn off or unplug the cooker at this point. Let it keep the rice warm for another 5 minutes before you unplug the rice cooker. When the rice is completely cooked, transfer the rice to a serving plate or bowl and carefully sprinkle both the "caramelized" onions and the roasted cashew nuts as garnish on the rice before serving.

Memorandums:

1. Let your guests know that all ingredients used in this recipe are edible, but it is generally recommended that they do not eat any of the cardamum pods, the cinnamon sticks, or the cloves.
2. In India, these ingredients are often eaten at the end of the meals as "breath mints" because of their strong pleasant smell.

<u>Serve 4-6</u>

Notes on these recipes from Dr. Chelliah:

1. Dr. Chelliah has given just 5 essential recipes which represent five different types of food.
2. These recipes are generic. For example, the beans recipe can be used for many different types of beans. By the same token, the lentil recipes can be used for many different lentils.
3. The purple cabbage recipe can also be used for green cabbage as well.
4. The spinach recipe can be used for collard greens, Swiss chard, or other greens.
5. The last two recipes are *for* high-carb low-fiber items. Eat them sparingly.
6. There are plenty of free vegan recipes on the internet.

DR. CHELLIAH'S WEIGHT LOSS PROGRAM

[Adapted from Dr. Greger's 21 Tweaks Program and Enhanced]

1. Walk 4 miles a day, or exercise for 1 hour every day upon waking up each morning before consuming any food.
2. Avoid All Animal Products.
3. Consume the following mixture every day:
    a) 6 tsp. of freshly ground flax
    b) Nutritional yeast 2 tsp.
    c) Ginger powder 1 tsp.
    d) Garlic powder 1/4 tsp.
    e) Black cumin 1/4 tsp.
    f) Cayenne pepper powder (red cayenne) ½ tsp.
    g) Cumin powder ½ tsp.
    h) Turmeric powder 1 tsp.*
    i) Black pepper powder 1/16 tsp.*
    j) Vinegar 2 tsp.
    k) Amla powder 1 tsp. (Asian Indian Gooseberry*)
    l) Low sodium tomato juice 4 ounces (add more if needed)
4. Mix thoroughly and drink every morning. If you wish, you can reduce the ingredients in half and consume it twice a day: for instance, before breakfast and supper.

5. Do this before each meal:
    a) Drink 2 glasses of water.
    b) Eat salad with dark greens and/or drink "hot" vegetable soup with no animal products or eat a ripe tomato.
    c) Consume diluted vinegar 2 tsp.
    d) Eat slowly: that is, a meal should be consumed over a period of at least 20 minutes.
6. Consume at least 3 cups of cold water brewed green tea grown in Japan: for low lead levels.
7. Consume a total of 4 quarts/liters of water daily (minimum 3 quarts/liters).
8. Consume at least 8 servings (1 serving = ½ cup) of dark greens daily.
9. Consume at least 4 servings of fully cooked beans daily.
10. Make sure all carbohydrates you consume have a carb: fiber ratio of 5:1 or less.
11. Do not consume any food after 7 PM until the following morning.
12. Sleep 7 hours each night.
13. Weigh yourself twice a day.

*There are no studies to support that *the asterisked items* cause weight loss, but they have many other benefits proven in controlled trials published in peer-reviewed journals.

DR. CHELLIAH'S CHIPS:

Ground Flax 2 pounds

Large red onions finely chopped 4

Ginger root peeled and finely chopped 4 inches

Garlic peeled and finely chopped 2 Bunches (not cloves)

Red bell (sweet) peppers cleaned of all seeds and finely chopped 4

Serrano peppers 4-8 (less if you don't like it spicy; more if you like it extra spicy) finely chopped

Bunches of fresh Cilantro leaves 2 finely chopped

Turmeric powder 2 tablespoons

Black pepper powder 1/2 tsp.

Nutritional yeast 4 tablespoons

Black cumin 1 tsp.

Cayenne pepper powder (red cayenne) 1 tsp.

Cumin (white) powder 1 tsp.

Cinnamon powder ½ tsp.

Cardamom powder ½ tsp.

Clove powder ½ tsp.

Amla powder 1 tsp. (Asian Indian Gooseberry*)

Salt 1 tablespoon

Water as needed

Mix all the ingredients thoroughly and spread the mixture onto oiled parchment paper and bake it at 170 degrees F for 36-48 hours. It should be dry when it is ready for consumption.

Note: These chips can be stored at room temperature for 2 weeks without refrigeration.

Part 11: Prioritizing Sweeteners for Foods and Drinks—From Sugar to Erythritol

A commonly prescribed healthy drink, in addition to carefully filtered water like Waiakea's Hawaiian deep well volcanic water that is naturally alkaline with electrolytes, is tea. Teas are numerous and largely beneficial, yet they can be ranked scientifically from most healthy to least beneficial. With some scientific support, this might be a workable ranking of teas: hibiscus, white, green, and black. Scientific findings suggest that we eliminate or minimize soft drinks, whether sugar free or not and that we eliminate or minimize all carbonated beverages. If we can, eliminate processed sugar. Replace sugar with sweeteners like erythritol and allulose—not aspartame, saccharin, stevia, monk fruit, or Splenda.

All substitute sweeteners for sugar that are currently available, except for erythritol and allulose, appear to be unhealthy. They appear to alter the microbiome: that is, the balance between the good bacteria and the bad bacteria in the gut (colon). When the balance is altered, instead of losing weight, we actually gain weight!! Although they are not artificial, xylitol and mannitol are osmotically active: that is, they act like Caster oil. They can result in diarrhea and flatulence. Technically, erythritol belongs to the same class, but it is the only one that is not osmotically active: that is, it does not cause diarrhea or abdominal bloating. Yes, it is true that xylitol, stevia, and monk fruit occur naturally, but they are still not good for us *because they can cause weight gain. Xylitol can cause* abdominal bloating and diarrhea. All no-calorie or low-calorie sweeteners available today, except for erythritol and allulose, can have some negative *health* effects. Dr. Chelliah stands by these statements on alternative sweeteners. He does not affirm that these sweeteners <u>will</u> cause *health* problems but that they <u>can</u> cause them.

Recent research has demonstrated that all no-calorie or low-calorie sweeteners such as stevia, aspartame, and monk fruit extract are dangerous. The only exceptions are erythritol and allulose. Allulose is an isomer of fructose which apparently has only 0.4 calories per gram (negligible) as compared to sugar which has 4 calories per gram. *All no-calorie or low-calorie sweeteners actually increased glucose and insulin levels in a delayed fashion.* Consequently, *they caused weight gain.* This elevation of glucose and insulin levels in the blood were missed because *they occurred several hours later following the next meal.* The original investigators were looking at the effect of *these no-calorie or low-calorie* sweeteners *on blood glucose* in the immediate hours after consumption comparing it to the effect of *table sugar on* blood glucose which goes up within the first hour after ingestion. Allulose and erythritol do not cause a blood glucose rise or an insulin rise *even in a delayed fashion.* In fact, allulose appears to improve insulin sensitivity! To applaud the value of erythritol, erythritol does not contribute to tooth decay, has antibacterial effects against streptococci bacteria, reduces dental plaque, does not raise blood sugar or blood insulin levels, and is an antioxidant.

We must rid ourselves of sayings that glorify sugar with sayings that reflect healthy forms of sweeteners. With Valentine's Day sweetness and humor intended behind these suggestions, transmute "sugar and spice and everything nice" into "allulose and spice and everything nice." Replace "you're even sweeter than sugar, honey" with "you're even sweeter than erythritol, allulose." When you can, oust "pretty please with sugar on it" for "pretty please with erythritol on it." In short, whenever you hear a saying ennobling table sugar, supplant it with the name for a healthy sugar substitute—erythritol or allulose will do fine. Fight the good fight against sugar with its nutritious superhero replacements. Nabal brags accordingly: "My friends think I am sweet enough. Nonetheless, I gobble up plain old tried and true white sugar."

Part 12: "Magic Mixes" for Health and Wellbeing

Diet can increase or decrease inflammation in the body. Consequently, carefully selected meals can be designed to be anti-inflammatory. Borrowing from Wikipedia.com, *anti-inflammatory* refers to a substance or treatment that reduces inflammation or swelling. In addition to medical drugs, some herbs and health supplements may have anti-inflammatory qualities: for example, bromelain from pineapples (Ananas comosus); Cannabichromene, a cannabinoid, also has anti-inflammatory effect; and, *St. John's wort's chief constituent, hyperforin, has been found to have an anti-inflammatory effect several times that of aspirin.* Diets that reduce inflammation include those rich in vegetables and low in simple carbohydrates and fats, such as saturated fats and trans fats. Anti-inflammatory foods may include most colorful fruits and vegetables, nuts, seeds, and certain spices, such as ginger, garlic, and cayenne.

Omega-3 fatty acids have been shown to reduce inflammation. This benefit however can be inhibited or even reversed if the ratio of Omega-6 to Omega-3 is too high as Omega-6 serves as a precursor to inflammatory chemicals (prostaglandin and leukotriene eicosanoids) in the body. The Dietary Inflammatory Index (DII) is a score (or a number) that describes the potential of a diet to modulate systemic inflammation within the body. *The DII can help dieters assess how anti-inflammatory their current eating choices are.*

Dr. Chelliah provides a healthy blend of vegan nutritional supplements and spices that help reduce inflammation, lower cholesterol, improve weight loss, and promote good health. His recipe, mix, or "magic potion" for overall health and heart health entails the following ingredients (Chelliah, 2022) mixed in water:

Dr. Chelliah's Magic Mixture

1. Nutritional yeast 2 tsp.
2. Ground flax 2 tsp.
3. Ginger powder 1 tsp.
4. Garlic powder 1/4 tsp.
5. Black cumin 1/4 tsp.
6. Turmeric powder 1 tsp.
7. Black pepper powder 1/8 tsp. (only a pinch)
8. Cumin powder ½ tsp.
9. Cayenne pepper powder (red cayenne) ½ tsp.
10. Amla powder 1 tsp.
11. Fenugreek powder 1/4 tsp.
12. Cinnamon powder 1/8 tsp. (a pinch or two)
13. Cardamom powder 1/8 tsp. (a pinch or two)
14. Clove powder 1/8 tsp. (a pinch or two)
15. Citric acid 1/8 tsp. (a pinch or two for taste to make the mixture palatable)
16. Salt for taste to make the mixture palatable and to balance citric acid

Practical vegans and vegans are advised to take this "magic potion" three times per day with meals. The mix can be sprinkled on almost anything that is eaten; it can also be stirred in a glass of water.

To lower cholesterol naturally without medications, Dr. Chelliah's ideal dietary program follows:

1 Avoid ALL animal products.

2 Have a handful of walnuts daily.

3 Drink one quart of alkaline water daily: *you can make your own (if alkaline water is not available in a local store)* with one quart of distilled water plus 3/4 of a teaspoon of baking soda.

4 Have three tablespoons daily of freshly ground flaxseed.

5 Consume eight servings of fully cooked beans or lentils daily.

6 Consume one teaspoon of amla powder daily.

7 Walk four miles a day (to raise HDL). Again, *substantial daily exercise complements his recommended healthful dietary implementations.*

8 Consume no more than 10 percent fat in food choices per day.

9 Eat eight servings of dark greens daily.

10 All carbohydrates consumed must have a CARB to FIBER ratio of 5 to 1 or less: that is, eat a high fiber diet.

Part 13: Practical Responses to Questions Frequently Asked about Health, Medicine, and Nutrition

The responses to health and nutrition questions are drawn from Dr. Chelliah's (2022) "Health Optimization" brochures. His health brochures are available by contacting Dr. Chelliah at drchelliah@heartnd.com or his office staff at 701-780-2000. We have chosen to select those brochures that would be especially relevant to practical veganism and veganism.

In his Health Optimization brochures, Dr. Chelliah cautions us *not to stop taking current medications without consulting a medical provider. He instructs us that his suggestions can be implemented in addition to medications being taken. Yet, he warns us to check with our medical provider before trying any of these nutritional strategies.* He is also aware that while he favors and recommends a total vegan diet for some conditions, he does not prescribe radical vegan eating habits on an all-or-nothing basis for everyone. To repeat, he practices non-invasive, invasive, and interventional cardiology. In fact, Dr. Chelliah qualifies his statements about meat eaters repeatedly in his writings on health through the phrase "on a regular basis." He warns those of us who eat fish, meat, chicken, and so forth that eating such meats "on a regular basis" will shorten our lifespan and healthspan. His ideal vegan diet consists of no meat or its products: no red meat, no white meat, no pork, no turkey, no chicken, no fish, no milk, no eggs, and no cheese. While Dr. Chelliah's practical vegan wholefoods plant-based diet only approaches his ideal vegan diet, it is still acceptable to him. As for Nabal, he testifies suitably: "I'm as healthy as the proverbial horse. I don't need some know-it-all doctor telling me that I can or can't take some medication or that I should or should not eat certain foods. I do what I want and take what I want and eat what I want when I want, and I'm as solid as the Rock of Gibraltar for doing things my way. Amen!"

With practical veganism, eating meats irregularly, especially when consumed occasionally and not daily, does not mean we have put ourselves in harm's way and will likely die soon. Dr. Chelliah does not demand that either we must always eat total vegan meals with even a trace element of meat removed or we will perish. When we refer in this text to eating or consuming specific meats, we do not mean that one large meal of chicken or beef will cause most people to die the next day. We contend that statements about eating and consuming foods harmful to us become harmful when eaten or consumed "on a regular basis" and notably in sizable portions like a 72-ounce steak several times a week—not merely one beef hot dog once a month. We are real and practical. Practical vegan diets that approach total veganism have Dr. Chelliah's realistic and practical support or this book would not have been written. As an emphatic reminder, *a practical vegan diet is one that embodies a high plant to meat ratio*: explicitly, not a PM ratio of 10 to 90 or 20 to 80—rather a PM ratio of 80 to 20 or 90 to 10.

Please consider and accept that Dr. Chelliah's recommendations are based on scientific findings and that 85 to 90 percent of all chronic diseases can be prevented or cured through a vegan diet combined with exercise. A practical vegan diet with a high ratio of plant to meat consumption would lean in the direction of attaining 80 to 90 percent of a total vegan diet. Even if the practical vegan diet falls short of the 100 percent perfect vegan diet, it still has enormous heart healthy benefits.

Let us set forth here and now some of the many health questions posed to Dr. Chelliah and his scientifically researched responses to these questions. For the convenience of the reader, in each categorical topic, the longer responses will be covered first and trail off to the shorter responses. His responses are grounded in peer reviewed articles in reputable medical journals. Please remain conscious of the fact that while Dr. Chelliah is a cardiologist and knows of conventional medical treatments through surgery and pharmaceuticals, he is choosing in the ensuing questions to limit his answers to non-invasive natural food sources to ward off or cure ailments. As he has said on many occasions, *when surgery is necessary due to the patient's cardiac condition, he orders the triple bypass or insertion of stents first and follows that procedure up with a prescription for a vegan or practical vegan diet.*

a Questions and responses pertaining to myths about veganism:

- *I hate vegetables. How can I eat what I hate? Why do most Norwegians enjoy lutefisk while most Americans do not?* Generally, we like the food we used to eat when growing up. In India and China, parents never have to order their children to eat their vegetables, for the children eat their vegetables first. Also, the way a food is prepared makes a big difference in its tastiness. Therefore, we can learn to like healthy foods that we disliked in yesteryear if it is cooked in a different manner. As you may recall, good food must first get past our lips to do us any good!

- *If we do not eat meat, how will we get enough protein?* There is plenty of protein in vegetable products: for example, beans and lentils. Besides, the daily requirement of protein for an adult is only approximately 0.36 gram for every pound of weight. With a body weight of 75 kg (165 pounds), a person should consume 60 grams of protein a day. That is only 60 grams of protein for the entire day for all three meals. Most of us eat far more protein than what our body requires. A whole-foods plant-based diet provides an adult athlete with the nutrition needed to be a winner.

- *Will I get enough energy to work hard or workout hard if I do not eat meat?* That muscles, strength, and endurance require the consumption of large quantities of animal-based foods is a myth. As mentioned earlier in this book, many Olympic champions have been vegetarians or vegans: namely, Charlene Wong (Canadian figure skater), Paavo Nurmi (Finnish runner), Carl Lewis (American track and field), Edwin Moses (American hurdler), and Murray Rose (Australian swimmer).

- *If we do not consume meat or milk, how will we get our daily requirement of calcium?* That vegetarians do not get enough calcium is a myth. Indeed, vegans and vegetarians need only about 500-700 mg of calcium each day which can be easily obtained by consuming a few servings of dark greens daily.

b Questions and responses pertaining to water:

- *What kind of water should we consume and how much do we need to consume daily?* Consume at least three quarts of water daily. While steam distilled water is best, avoid soft water. Alkaline water is better than pure water that is not alkaline. Avoid acidic water entirely. Water can be made alkaline by adding 3/4 of a teaspoon (tsp.) of baking soda to a quart of distilled water. Nonetheless, Dr. Chelliah advises us to check with our doctor before trying this.

- *Is bottled water safer than tap water?* Not necessarily. Some bottled water has been found to be contaminated with bacteria.

c **Questions and responses pertaining to fruit juices (and fruit) and vegetable juices (and vegetables):**

- *Is there any way to boost the nutritional value of low salt vegetable juice in a can?* Although Dr. Chelliah does not recommend commercially produced vegetable juice, if commercially produced vegetable juice is the only available option, we can improve it by adding this recipe: take one can of low salt V8 vegetable juice + 1/4 tsp. of turmeric + one pinch of black pepper + 1/4 tsp. of Indian gooseberry powder + 1/4 tsp. of fenugreek powder + 1/8 tsp. of lemon juice + a pinch of salt + one tablespoon of nutritional yeast.

- *What fruit juice is best for us?* Avoid all fruit juices and consume unprocessed fruit (as long as we are not diabetic and are not trying to lose weight). Although fruit is healthy, regular consumption of fruit will prevent us from losing weight. If we are at our target weight and do not have diabetes, we may eat as much fruit as we like.

- *Are vegetable juices healthy?* Fresh (or frozen) vegetables are the best for us. The second best is vegetable juice we make for ourselves using *whole* vegetables. The third best involves the consumption of commercially produced vegetable juice (as long as it does not contain a lot of salt or preservatives).

- *Is pomegranate juice beneficial?* A study sponsored and funded by the pomegranate industry failed to show any benefit.

d **Questions and responses pertaining to healthy hot chocolate, coffee, tea, and alcohol:**

- *Is hot chocolate healthy for us to drink?* It can be if this recipe is *used*: 1 glass of soy milk + 1 tsp. of pure unsweetened cocoa powder + erythritol. Erythritol is derived from fruit and is a natural no-calorie sweetener that is also an antioxidant. Let us note that cocoa is a powerful antioxidant by itself, and it has a tiny amount of caffeine that is clinically insignificant.

- *Do you recommend coffee or tea?* White decaffeinated tea is a powerful antioxidant. Adding a bit of lemon gives us increased chemical benefits. A recipe for hot or cold tea follows: One quart of steam distilled water + 2 tablespoons of erythritol + 2 to 3 packets of white and/or green decaffeinated tea. Steep for 15 minutes. Add 1 tsp. of lemon juice.

- *What is your assessment of caffeine?* Although caffeine has some beneficial properties, especially when consumed in small amounts, the disadvantages outweigh the benefits. Caffeine consumption is associated with increased incidence of peptic ulcer disease, palpitations, nervousness, and high blood pressure.

- *How much alcohol might be good for us?* Alcohol should be avoided entirely. Although purportedly it has some benefits, the risks far outweigh any benefits. Even rinsing our mouth with alcohol containing mouthwash increases the risk of cancer. Even a tablespoon of alcohol kept in the mouth for 10 seconds and spat out increased the risk of breast cancer.

- *Are herbal teas OK?* Most herbal teas are OK. Hibiscus tea is useful in the treatment of hypertension (high blood pressure). However, if you develop a dry cough, please discontinue it.

e Questions and responses pertaining to oils:

- *Is flaxseed oil better than fish oil?* First of all, *ground flaxseed is better than flaxseed oil.* Of course, flaxseed oil is better than fish oil. Fish oil should be avoided because commercially produced fish oil is contaminated with dioxins and PCB toxins.

- *Which Omega-3 fatty acid supplements do you recommend?* DHA supplements are useful provided they are derived from plant sources like algae. An affordable commercially sold algae derived Omega-3 supplement is available from Nature Made Nutritional Products.

- *What is the most nutritious oil for us to consume?* The best oil for consumption is the oil that is still within the nut or the fruit. For example, if you want the benefit of olive oil, *eat the olives*; if you want the benefit of walnut oil, *eat the walnuts*. However, *the best oil for cooking is algae oil* which is only available from one manufacturer: namely, *Thrive*. The second-best oil is the high oleic acid sunflower oil.

f Questions and responses pertaining to vegetables and nuts:

- *Which vegetables, beans, and grains should we consume?* Consume one or more of the following vegetables daily for their health benefits: Kale (anticancer plus calcium and iron), spinach (anticancer plus calcium and iron), broccoli (anticancer plus calcium and iron), cauliflower (anticancer), and white mushrooms (anticancer). Consume one or more of the following beans and grain daily: black eye beans, red beans, soybeans, mung beans, garbanzo beans, any other type of beans, and of course lentils. Nabal pooh-poohs the idea of beans. He says: "There's absolutely, positively no way around feeling bloated and gassy. And, that's that!"

Dr. Chelliah explains, however, that special conditions for consuming beans can be implemented:

First, *gas in the stomach from consuming beans can be greatly reduced or eliminated by* sprouting the beans, simply soaking the dry beans in water overnight with one teaspoon of baking soda added to the water or by consuming 1/8 of a tsp. of Asafoetida powder with an ounce of water just before eating the beans.

Second, do not trust the dietician who declares that diabetics should not eat beans. *As long as beans are cooked without sugar, molasses, or meat, beans constitute the best sources of protein and other healthy nutrients.* The high fiber content of beans blocks the blood sugar raising effects of carbohydrates in the beans. To illustrate, if we have a carbohydrate serving containing 10 grams (g) of "sugar" and that same serving of carbohydrate contains 10 g of fiber, we consume zero ("O") carbs! Thus, *one gram of fiber cancels one gram of carbohydrate.*

Third, avoid diets that recommend high protein or high fat/low carbs to assist us in losing weight. High fat/low carb diets are dangerous and can lead to ketosis, bad breadth, and - at the risk of sounding dramatic - even death. However, *Dr. Chelliah recommends that type 2 diabetics and even others who would like to improve their overall health limit the daily consumption of "NET carbs" to about 40 grams per day.*

- *Do you have anything to say about purple cabbage, purple potatoes, purple tomatoes, mushrooms, hot chili peppers, and spirulina?* The best type of cabbage is purple. The most nutritious potatoes and tomatoes are purple. Raw mushrooms are not good for us. There is a toxin, agaritine, in raw mushrooms that is destroyed by cooking. So, we should eat a lot of cooked mushrooms—not raw mushrooms. Again, perhaps counterintuitively, *the idea that raw food is always better than cooked food for us is wrong.* White mushrooms can prevent cancer. Chanterelle mushrooms are the most nutritious followed by white mushrooms. Furthermore, white mushrooms can lower our risk of cancer. As for hot chili peppers, in general, they are not harmful and are good for us. However, spirulina contains contaminants which can cause cancer and therefore are not good for us.

- *What about sprouts, calcium, tomatoes, and frozen vegetables?* Avoid alfalfa sprouts, and eat broccoli sprouts instead. The *best sources of calcium* are green leafy vegetables like kale and spinach. *Tomatoes contribute to prostate health.* Perhaps counterintuitively, *cooked tomatoes are far more nutritious than raw tomatoes.* Cooking tomatoes releases four times more lycopene, which might yield health benefits, from the tomatoes. While *fresh vegetables* are best for consumption, *frozen vegetables* embody an acceptable alternative since frozen vegetables lose only a small percentage of nutrients.

- *Can you tell us something about vegetables and cancer?* To reduce our chances of developing cancer, *no one vegetable decreases the risk of all cancers.* For example, while radishes are effective against stomach cancer, they are not effective against pancreatic cancer. Similarly, although orange bell peppers can cut prostate cancer growth by more than 75 percent, they are ineffective against stomach cancer. In short, *because each vegetable tends to attack different cancers, we should eat a variety of vegetables.*

- *What can you tell us about healthy nuts?* Although nuts have high fat content, generally speaking, *a handful of nuts consumed daily can*: improve cholesterol levels and improve male sexual function (pistachios in particular). From most to least beneficial for our health, nuts can be ranked in this order: pecans, walnuts, hazel nuts, pistachios, and almonds with pine nuts as the least nutritious of nuts.

g Questions and responses pertaining to spices, vitamins, and herbs:

- *Can you enlighten us on what to do with spices, vitamins, and herbs?* Indian Asian gooseberry (Amla) powder is the most powerful antioxidant discovered to date. It is also a powerful anti-diabetes substance. *Amla powder lowers LDL cholesterol by 20 to 40 percent and raises HDL by 15 to 20 percent.* It is an anticancer agent and prevents skin wrinkles.

A dose of one teaspoon daily is recommended by Dr. Chelliah's research findings. *Fenugreek* powder is another anti-diabetes substance. It improves muscle strength and adds fiber to our diet. One teaspoon daily is Dr. Chelliah's prescription. Taken daily, *saffron* can be effective in treating depression. A daily dose of 30 mg of saffron was as effective as the prescription medication Aricept for treating Alzheimer's disease. *Black pepper* has anti-inflammatory and anticancer effects. *Turmeric* may reduce inflammation that Dr. Chelliah says can result from as little as 1/4 of a teaspoon daily. He adds that turmeric benefits can be sizably enhanced by as much as 2,000 percent by simply spiking the 1/4 teaspoon with a pinch of pepper.

With respect to vitamins, Dr. Chelliah maintains that most people eating a balanced vegan diet do not need vitamin supplements except for daily doses of vitamin B12 (50 micrograms—mcg—daily), vitamin D (2,000 to 3,000 units per day), and DHA (250 mcg of Omega-3) from a plant source like algae, not fish. The *optimal dose of Vitamin D supplements* has been determined: *less* than 2,000 units per day is inadequate while more than 3,000 units per day may be harmful. Vitamin A and E supplements may harm most of us unless we suffer from a vitamin A or E deficiency.

- *What can Dr. Chelliah tell us about other spices and herbs?* Depending on which cinnamon we consume, cinnamon can hurt us or help us. Chinese cinnamon or Cassia has a toxin. Ceylon cinnamon or true cinnamon may have some beneficial properties when used in small amounts. In brief, we should avoid Chinese cinnamon while using Ceylon cinnamon with care: *in small amounts*, if we really like it, yet not just for the benefit—a benefit that currently shows no clear advantages. Although nicknamed the "stinking gum," *asafoetida* has anti-flatulent and antiviral properties and may accelerate the healing of a wound. There is some scientific evidence that chemicals in asafoetida might help treat irritable bowel syndrome (IBS) and might protect us against high blood levels of certain fats including cholesterol and triglycerides. Chemicals in asafoetida can thin the blood. For other healthy options, we can ingest oregano, marjoram, allspice, dried cloves, lemon balm, dried peppermint, and ginger which enhance our health. Yes, they are all good for us. *Marjoram is in the oregano family, yet it is more than 50 percent more powerful than oregano.* Dried lemon balm makes good tea. Grated ginger is effective against nausea and may protect us from the harmful effects of dioxin and other industrial toxins like DDT.

Ginger deserves special attention. Airplane travelers, airline pilots, airline flight attendants, workers in hospital X-ray departments, and everyone subjected to radiation risk chromosomal damage and oxidative stress to their bodies. In one study, researchers compared the oral intake of ginger to the leading drug injected into people for radiation sickness. They found ginger to be 150 times more powerful than the drug without the serious side effects of the drug. If we were to take a tissue sample from someone, about 7 percent of that person's cells may evidence DNA damage. However, just including ginger in that person's diet may cut that damage by 25 percent. *Ginger* thus constitutes an inexpensive natural product that *may protect us against radiation-induced DNA damage.* So, eating crystallized ginger *(made with erythritol instead of sugar)* when flying may not only prevent travel sickness but may also protect us from the cosmic radiation at high altitudes.

In addition, nausea and vomiting are common during early stages of pregnancy, affecting up to 85 percent of women worldwide. The best available evidence suggests that *ginger* is a safe and effective treatment for women who suffer from nausea and vomiting during the early stages of pregnancy. The recommended dose is a gram of powdered ginger per day: or about half a teaspoon of grated fresh ginger or four cups of ginger tea. For pregnant women, the maximum recommended daily dose is four grams per day. Higher doses may induce uterine contractions.

Ginger also helps in the management of postoperative nausea and vomiting and in preventing anti-retroviral induced nausea and vomiting during HIV treatment and chemotherapy induced vomiting. Nabal defies this statement: "Oh, yeah, show me the studies if you have any, Mr. Big Shot doctor." One randomized, double-blind, placebo-controlled clinical trial of ginger for the management of vomiting caused by chemotherapy for breast cancer found that vomiting was relieved in all phases of chemotherapy. In other words, in the acute phase within 24 hours of the chemotherapy, ginger relieved delayed vomiting two to three days later. It was even beneficial in the management of anticipatory vomiting—occurring before a chemotherapy session. In patients receiving a series of chemotherapy sessions, after the first few sessions, the body knows what is coming and starts throwing up at the thought of the approaching chemotherapy session. Costly standard drugs do not seem to control anticipatory nausea. Ginger costs about two pennies per dose and works.

With respect to PMS in women, painful periods are common and can cause severe suffering. For alleviating menstrual cramps, ginger may also be as effective as ibuprofen. Four randomized controlled trials published have evidenced that *ginger can alleviate menstrual pain*. All four studies showed significant benefits when taken the first few days of a period. An effective dose ranged from a third to a full teaspoon a day. Since all doses worked about the same, it is advised to start with a low dose. As a side benefit, ginger can dramatically reduce heavy flow—one of the most common gynecological problems for young women.

For the 18 million young women in the United States who experience iron deficiency anemia due to heavy menstrual bleeding, reducing blood loss is beneficial. In one study, the blood loss was decreased by 50 percent by taking merely an eighth of a teaspoon of *ginger powder* three times a day during the menstrual period. The dose was found to be effective for menstrual cramps as well as for the reduction of blood loss. *Ginger helped manage premenstrual syndrome* as well by taking an eighth of a teaspoon of ginger powder twice a day for a week before the period. Ginger yielded a significant drop in PMS mood, physical symptoms, and behavioral symptoms. By the way, dietary interventions *reducing salt and animal fat* intake are also favorable for lessening PMS symptoms.

Ginger can mollify migraines. A double-blind randomized controlled clinical trial comparing the efficacy of ginger to that of the drug sumatriptan (known as Imitrex) for the treatment of migraine headaches showed that an eighth of a teaspoon of powdered ginger was as effective as the standard dose of the drug. Both worked just as well and fast. Most suffering from migraines started out in moderate or severe pain before taking ginger or the drug. After taking ginger or the drug, those with migraines suffered but mild pain or were pain free. *The same proportion of migraine sufferers in the ginger and drug groups reported satisfaction with the results.* Also, in the ginger group, there were fewer side effects and ginger is a lot less expensive than the drug. In short, ginger won.

In an osteoarthritis study published in 2001, ginger was compared with a placebo for managing osteoarthritis. It proved to be only marginally superior to the placebo. However, *subsequent research showed ginger to reduce pain and disability in patients with osteoarthritis when compared with conventional anti-inflammatory drugs like ibuprofen.* Moreover, anti-inflammatory drugs can have harmful cardio-vascular and gastro-intestinal side effects. Anti-inflammatory drugs can cause stomach irritation and bleeding. Pointedly, in one study, ginger consumption increased stomach lining protection. Noteworthy, at a dose of 1/4 to 1/2 a teaspoon per day, ginger may be consumed with drugs like ibuprofen to reduce the negative side effects of ibuprofen.

h Questions and responses pertaining to food additives, longevity, and diverse topics:

- *Are there any steps we can take to prolong our lives and health?* Recent research has discovered that *calorie restriction, reducing the amount of calories consumed daily, may extend our lifespan.* Methionine, an essential amino acid, promotes cancer growth when consumed in excess. *Methionine restriction in the diet can achieve lifespan extension just as much as calorie restriction.* Animal proteins typically have a high methionine content whereas plant proteins are low in methionine. Many cancer cells are dependent on methionine for their growth whereas normal cells can grow even when the methionine level is low. Resveratrol derived from purple grapes may also be beneficial in extending life but unfortunately may cause kidney failure.

  Consumption of beans also prolongs life. Consumption of just 20 grams (two tablespoons) of beans per day reduces the risk of dying by 8 percent.

  A longevity account may be worth mentioning here. The bristle cone pine is the oldest living tree on earth. There is one in California that started growing about 5,000 years ago and is still growing. Scientists found an enzyme in its roots called telomerase. Humans also have telomerase. Telomerase rebuilds telomeres. Telomeres are end caps on our DNA. Telomeres wear down like a pencil eraser each time the cell divides. When the telomeres are completely gone, the cells stop dividing. Old cells are not replaced by new ones because no new cell can be produced. Thus, telomeres operate as a kind of "life fuse." *Telomere fuses start shortening from when we are born. When these fuses are gone, so are we.* Forensic scientists use this fact to estimate the age of someone by examining the length of that person's telomeres. *The shorter the telomeres, the older the person is.*

The wishful thought is this. If we can slow down this shortening of the telomeres that is like a ticking timebomb, we may be able to slow down aging and live longer. Is this merely a hopeful idea with a hopeless future? Actually, there are dietary options that slow down the shortening of the telomeres. *Smokers should stop smoking because smoking shortens the telomeres.* Two types of foods can be eliminated or minimized since they are associated with telomere shortening and therefore accelerated aging in humans: fish and processed meat. Consuming fish "on a regular basis" appears to shorten the telomere which shortens human life by around 6 years. Consuming processed meats "on a regular basis" appears to shorten the telomere which shortens human life by 14 years.

The dietary solution is implicit in these findings. To slow down the aging process and optimistically even turn back the clock of aging can be achieved through the adoption of a wholefoods plant-based vegan diet. The shortening telomeres problem has only recently found a solution. In a study funded by the U.S. Department of Defense, Dr. Dean Ornish found that *after only three months on a wholefoods plant-based diet with exercise, telomerase activity was boosted significantly.*

How else can we extend our lives? While caloric restrictions can be challenging and not very practical, studies have shown that caloric restrictions can significantly extend our lifespan. Most people cannot follow calorie restricted diets because they feel intolerably hungry as a rule. As Dr. Chelliah has suggested, adding beans and lentils to the low caloric diet might ward off hunger and make the low-calorie diet workable. However, there may be a better way, according to Dr. Chelliah. Scientists discovered that the benefits of dietary restriction may be coming not from restricting calories, but from restricting protein intake. *Just reducing protein intake without any changes in calorie intake has effects similar to restricting calories.* In fact, protein restriction is easier to maintain than calorie restriction and may even be more effective than calorie restriction in extending our lifespan. If only the amount of amino acid leucine consumed is reduced, it might be just as effective as cutting down on proteins. In general, to lower leucine levels, we should restrict animal proteins.

Furthermore, to extend life, Dr. Chelliah calls our attention to methionine which promotes cancer. Restricting methionine may very well result in expanding lifespan. Since plant-based diets tend to have low levels of methionine while animal-based diets tend to have high levels of methionine, plant-based diets would be advantageous in terms of lifespan extension over animal-based diets. Expressly, *a vegan or practical vegan diet, when combined with exercise, is the safest way to prolong life. Three clinical narratives on vegan diets and exercise come next.*

An ironic account emerges from the harshness and savagery of Nazi prison camps in WWII. Note that coronary blood vessels are the tiny blood vessels which supply blood and oxygen to the heart muscle. When a prisoner died, an autopsy was done routinely. Ironically and counterintuitively, *regardless of age, not a single blockage was found in the coronary blood vessels of any of the prisoners.* In pathetic contrast, during the Gulf War, *when soldiers ranging in age from 19 to 24 died and autopsies were done, those conducting the autopsies found several blockages in the coronary blood vessels of the young soldiers.* Why is there such a contrast between elderly prisoners with no blockages and young soldiers with blockages? *The prisoners in the Nazi prison camps ate very little high fat animal products whereas the young soldiers in the Gulf War ate high fat animal products.* Hence, food made the difference.

A comeback story follows. Experiencing chest pains, inventor and longevity researcher Nathan Pritikin went to his physician and underwent a stress test which was strongly positive (abnormal). His physician told him he needed to go home and take it easy and that not much could be done for his coronary artery disease. He received pain medications and was advised to live out the rest of his life to the best of his ability and to prepare to die soon. *Not liking what he heard, Pritikin took matters into his own hands and went on an extremely low-fat diet with small amounts of meat.* Within a few weeks, he noticed a sharp *decrease in chest pains* and felt better. He *began to exercise and experienced additional improvement.* Impressed with his results, Pritikin wrote books about his experiments on himself. It has been reported that based on his request before his death, after his death, an autopsy was performed. The autopsy revealed impressively that his coronary vessels were found to be clean as reported in the *Los Angeles Times* newspaper on 4 July 1985.

An uplifting history ends this clinical trilogy. *National Geographic* magazine published an article on 16 November 2005 of a 30-year study of the three groups of people in the world who lived the longest: Okinawans in Japan, Sardinians in Italy, and Seventh Day Adventists in the United States. The researchers discovered that during the last 10 years of the study, only the Adventists continued to demonstrate an enhanced lifespan. The reasons for this groups continuing longevity seems to point to these facts: 1) While 50 percent of Adventists are vegetarian, the other 50 percent eat meat occasionally or rarely; 2) Adventists abstain from alcohol, tobacco, caffeinated beverages, do their best to exercise, and sleep 7 hours every night. The incidence of all diseases (including cancer, heart attacks, strokes, kidney failure, and infection) in this multiracial group was lower, and they lived an average of between 5 and 7 years longer than the average American population. *The Adventists had longer lifespans and healthspans.* The most important point to consider here is that their quality of life was better. *We Americans tend to live longer now than before, but most elderly Americans also are not enjoying the last few years of their lives because they suffer from chronic illnesses.*

- *What can we do to avoid kidney stones?* Calcium oxalate (sometimes combined 80 percent with calcium phosphate) is the most common type of kidney stone. To reduce the risk of kidney stones, Dr. Chelliah suggests the following steps:

  1) Drink plenty of water (two quarts per day).
  2) Avoid grapefruit juice, apple juice, and tea.
  3) Since vegetarians have 50 percent fewer kidney stones than meat eaters, avoid all animal products. Note well: *animal protein consumption* is associated with hypercalciuria (too much calcium in the urine), hyperoxaluria (too much oxalate in the urine), hyperuricosuria (too much uric acid in the urine), hypocitraturia (much less citrate in the urine), and the acidification of the urine.
  4) Avoid consuming high doses of vitamin C.

5) Avoid or reduce the consumption of dark greens which contain increased amounts of oxalate. Avoid greens such as spinach. *Arugula, romaine lettuce, kale, mustard, turnip and collard greens, bok choy and watercress* contain a much lower amount of oxalate.

6) Avoid soy products, almonds, potatoes, beets, navy beans, raspberries, and dates.

- *But what about breast cancer?* How can diet help us survive breast cancer? Breast cancer survival can be dramatically improved by four times by simply adding *daily exercise* and including *five servings of fruit and vegetables*, especially cruciferous vegetables, daily. According to Dr. Chelliah, daily exercise with a vegan diet may be almost as effective as many chemotherapy regimens for preventing the recurrence of breast cancer in some women. Because Dr. Chelliah knows there is always a place for modern chemotherapy in patients with breast cancer, he advises all women with breast cancer to consult their cancer specialist. In reducing the risk of breast cancer, we can also help ourselves by taking 3 teaspoons of ground flaxseed daily and consuming broccoli and white mushrooms. *Of course, if breast cancer is detected, Dr. Chelliah would support a treatment with all appropriate modern surgical, radiation, and chemotherapy along with a vegan diet.*

More needs to be said about avoiding and treating breast cancer. One in eight American women will be diagnosed with breast cancer in their lifetime. The current medical approach to treating cancer consists of chemotherapeutic drugs and radiation. There is one relatively new therapy called *anti-angiogenesis* that aims to cut off the blood supply to tumors. When tumors start to grow, they need blood. Therefore, tumors do their best to form new blood vessels to feed themselves—called *angiogenesis*. Medical strategies designed to prevent the blood vessels from growing in a tumor is called *anti-angiogenesis*. Another strategy is to strike at the root of the cancer—namely, the cancer stem cell. The strategy of striking at the cancer stem cell does not require an expensive new drug; it only requires plain and lowly broccoli. Sulforaphane, a dietary component of broccoli and broccoli sprouts, appears to inhibit breast cancer stem cells, and a vegan diet may suppress angiogenesis.

In one study, after eating a lot of soy for two years after surviving breast cancer, survivors eating a lot of soy were still alive. However, a quarter to a third of the women who ate the *least* soy had died. After five years, 90 percent of the women who consumed soy products were still alive while only 50 percent who did not consume soy products were still alive. In summary, it appears that *soy intake is associated with increased survival and decreased recurrence of cancer.* The idea that soy products promote breast cancer is patently false.

Several diverse dietary ways to reduce breast cancer are noteworthy. A study from Germany found that reduced breast cancer risk was associated with those who ate sunflower and pumpkin seeds. Other studies have additionally revealed that people who eat legumes—beans, chickpeas, split peas, and lentils—live longer. Also, another option for treating breast cancer is flaxseeds, and flaxseeds contain lignans. *Some breast cancer cells thrive on estrogen.* If there are not enough estrogens in the body, the breast cancer cells do something very clever. They produce aromatase which is an enzyme that can convert the small amounts of androgens in women into estrogens. *(Both men and women have both androgens and estrogens, but women have more estrogens than androgens and men have more androgens than estrogens.)* And, of course, a common and inexpensive cooked button mushroom has chemicals which can block or destroy *by up to 65%* aromatase—an enzyme that converts androgens to estrogens which breast cancers thrive on. These chemicals in white mushrooms appear to be unsaturated fatty acids such as linoleic acid, linolenic acid, and conjugated linoleic acid. But Dr. Chelliah cautions us: "Never substitute an 'active' ingredient for the whole food." So, eat the cooked mushrooms, not any mushroom extract! *Remember that raw mushrooms should be cooked to destroy the poison agaritine which is quickly killed when cooked or heated.*

There is further evidence that cholesterol may contribute to the development and progression of breast cancer. *Cancer feeds especially on LDL cholesterol:* that is, the "bad" cholesterol. A large study found that breast cancer in women increased by 17 percent if their cholesterol was over 240, contrasted with women whose cholesterol was under 160. Also, lowering cholesterol also has the benefit of reducing the number of heart attacks and strokes.

A summary of the wholefood plant-based strategies for women to reduce their cancer risk of breast onset and recurrence include:

1) Avoid all animal products (to lower IGF-1 levels).
2) Consume a low-fat wholefoods plant-based diet.
3) Eat broccoli several times a week.
4) Consume soy products.
5) Eat beans and lentils.
6) Eat cooked mushrooms.
7) Consume freshly ground flax seed.
8) Eat sunflower and pumpkin seeds.

- *Are any of these food additives safe: sodium Benzoate, artificial colors, carrageenan, special fruit juices, and salt?* The answer for these additives is "no"—*not one is safe.* When sodium benzoate is combined with ascorbic acid (vitamin C), as it is often done in soft drinks, it can become carcinogenic. Artificial colors can cause cancer. Over certain consumption levels, carrageenan can be toxic. A special fruit juice like mangosteen contains xanthones. Xanthones poison the cellular metabolism. As for salt, *even if we have a normal blood pressure (BP), we should use salt sparingly.*

- *What can we do to lower muscle soreness after exercising?* Muscle soreness after exercise is caused by inflammation of the muscle. The delayed onset of muscle soreness that follows in the days after extreme physical exertion is most likely due to inflammation resulting from muscle cell damage: that is, from little microtears in the muscle. The anthocyanin flavonoids in *berries* may help relieve the soreness. Especially helpful are *cherries and blueberries.*

- *Is there anything we can eat to burn and get rid of fat?* Eat foods containing concentrations of arginine. Regarding arginine supplements as pills or capsules, Dr. Chelliah does not recommend arginine in supplemental form. Rather, he endorses *arginine from these natural sources*: pumpkin seeds, squash seeds, watermelon seeds, sesame seeds, peanuts, soybeans, peanut butter, tahini, almonds, pine nuts, fava beans, and sunflower seeds.

- *Is there anything we can do to improve eye health?* The most protective dietary component for *decreasing the odds of glaucoma* by 69 percent is consuming *at least one serving a month of collard greens or kale*. A vegan wholefoods and fruit diet can significantly slow down the onset of *age-related macular degeneration* (ARMD).

- *Is there anything we can do to improve our exercise endurance*? We can *drink beet juice*. Drinking beet juice is associated with greater endurance with less oxygen consumption. With less oxygen required, drinking beet juice resulted in a 16 percent increase in endurance in a beet juice group in a study. Nabal asks: "Is beet juice the same as beetle juice? In the movie *Beetlejuice*, did Michael Keaton drink beetle juice or beet juice? I'm a little confused. Help me out here. I want to have more endurance for exercise."

- *What can we do to reduce muscle fatigue*? To reduce muscle fatigue and the burning sensation in the muscles that lactic acid accumulation causes, *eat an orange before exercising*.

- *For those of us with Inflammatory Bowel Disease (typically, Crohn's disease and ulcerative colitis), what can we do to stay in remission?* Ultimately, avoid all meats and become a vegan or vegetarian. A study done in Japan revealed a remission rate of 100 percent at one year and 92 percent remission at two years. *People with inflammatory bowel disease should avoid nutritional yeast.*

- *What can be done to prevent Parkinson Syndrome?* The following health strategies might be helpful:

    1) Exercise with reasonable vigor.
    2) Avoid dairy products, for toxins like tetrahydroisoquinoline occur predominantly in cheese.
    3) Avoid pesticides, such as PCBs (138, 153, and 180), all of which are found in meat.
    4) Avoid head trauma.

- *What can we do to strengthen our immune system?* The recommendation from Dr. Chelliah is a teaspoon of nutritional yeast. A teaspoon *or even less* of *nutritional yeast* can reduce the rate of infections by 50 percent by *boosting immunity*.

- *Are there any dietary strategies for controlling multiple sclerosis when the disease is early in its development?* One study showed that 95 percent of patients who eliminated saturated fats (animal products) remained asymptomatic for 34 years. Recently, there is some new evidence that Epstein-Barr virus may be implicated, yet diet modification is still a good strategy.

- *What foods can help with my seasonal allergies?* Four types of foods can contribute to lessening seasonal allergies: seaweed (except hijiki and kelp), dark green leafy vegetables, flax, and miso.

- *Are there any natural therapies for noninsulin dependent diabetes?* Two are prescribed: the Indian gooseberry powder (also known as Amla powder) and fenugreek powder.

- *Can we eat butter flavored microwave popcorn without fear of health problems?* Sadly, increased consumption of butter flavoring causes a type of lung disease—Bronchiolitis Obliterans—that can be fatal.

- *Are there natural foods that can improve erectile dysfunction (ED)?* ED can be improved by eating pistachios and watermelon (especially yellow watermelon).

- *What is an effective treatment for constipation? Prunes* are better than Metamucil!

i Questions and responses pertaining to animal products:

- *What is the downside of milk consumption?* Milk consumption is associated with acne. Casomorphins liberated from the cow's milk protein beta-casein have been implicated (although not yet proven) in the cause of such conditions as: autism, crib death, type I diabetes, postpartum psychosis, and food allergies. Milk protein casein is converted in the stomach to casomorphin which stimulates the opioid receptors in the mucus glands of the respiratory system to produce excess mucus. This might be the reason why many patients with asthma improve dramatically when they stop consuming milk. All milk products contain casein, a protein, which stimulates *a hormone called IGF-1. IGF-1 causes rapid growth of all cancer cells.*

IGF-1 stands for Insulin like Growth Factor 1. This hormone is produced primarily by the liver in response to the growth factor that is released from the pituitary gland inside the brain. Casein (milk protein) is a powerful stimulator of IGF-1. IGF-1 is a powerful stimulator of cell growth, but IGF-1 does not discriminate between normal cells and cancer cells. *Since cancer cells naturally multiply much more rapidly to begin with, IGF-1 makes these cancer cells grow uncontrollably.*

Scientists at the Pritikin Institute in New York did research relating to IGF-1. Nabal protests boldly: "Research institutions are questionable. How can anyone trust them? A bunch of hoity-toity number- crunchers thinking they are smarter than God! Why should I believe and trust a bunch of overpaid snoopers who are only interested in getting government funding for their pet projects that are totally biased and corrupt?" Well, the Pritikin researchers measured the IGF-1 in meat eaters and vegans. The scientists found that *IGF-1 levels were much higher in meat eaters than in vegans*. They took a drop of blood from meat eaters and dropped it onto cancer cells growing in a Petri dish and observed that the blood of meat eaters killed about 9 percent of the cancer cells. Repeating the procedure with a drop of blood from vegans, the blood of vegans killed 8 times more cancer cells—that is, 72 percent. Next, they put the meat eaters on a vegan diet for just over 10 days. When they measured the IGF-1 level in their blood, the IGF-1 level had dramatically dropped to the same level as that of vegans. The scientists then took a drop of blood from the former meat eaters who became vegans for just under two weeks and dropped it on the cancer cells. They found that their blood killed cancer cells 8 times better than their previous blood sample while on a meat diet—the same as the blood of lifelong vegans. The scientists demonstrated that the cancer cell "kill" rate improves in just two weeks after switching to a vegan diet. In the last part of their experiment, when the scientists added back to the Petri dish containing cancer cells *and* the blood of former meat eaters who became vegans, enough IGF-1 to return the IGF-1 level to that of meat eaters, the cancer cell "kill" rate dropped back to 9 percent. The experiment supported the idea that *IGF-1 contributes to cancer cell growth.*

An epidemiological clinical tale about IGF-1 is worth reporting. Laron Syndrome is a rare disease in which people are born with the inability to respond to IGF-1 growth hormone:

that is, they don't have enough "receptors" to respond to the hormone. Since the IGF-1 growth hormone provides for normal development in humans, the lack of the IGF-1 growth hormone receptors leads to people very short in stature with a small body frame—in other words, "little people." A family of 99 in Ecuador were born with birth defects associated with a deficiency in the IGF-1 growth hormone receptors. Although those born with the birth defect had short stature and small frames, their cousins with normal levels of the growth hormone had normal statures. However, the cousins with "normal" levels of the IGF-1growth hormone receptors had about the same percentage of cancer development and cancer related death as the general population—around 20 percent. In startling contrast, their cousins with the IGF-1 growth hormone receptor deficiency had zero ("0") percent cancer and diabetes.

The *consumption of animal products is associated with higher levels of IGF-1 and a higher incidence of cancer than the frequency of cancer in vegans.* Moreover, even after one develops cancer, switching to a vegan lifestyle dramatically improves the cancer cell "kill" rate. Increasing the cancer cell kill rate may even reduce the cancer victim's chance of cancer returning.

- *What can you tell us about melatonin and breast cancer?* Recent research reveals an inverse correlation between melatonin levels and breast cancer risk. The pineal gland in the brain produces melatonin. The pineal gland is active only when there is no light around our surroundings: that is, when you are sleeping in a dark room. The pineal gland stops secreting melatonin once light hits our eyelids. As a reminder of content covered earlier in this book, researchers confirmed this when they were able to document that blind women have 50 percent lower rates of breast cancer than women with full eyesight. Furthermore, all *animal products suppress melatonin levels whereas plant products do not.* Consequently, Dr. Chelliah can affirm that *adequate sleep is vital in preventing cancers as well as other diseases*: such as obesity, diabetes type 2, high blood pressure, heart attacks, and cancer. Thus, a sleeping room should be completely dark to keep out all light. A plain clothespin can be used to pin drapes together to block out all light.

- *How about pork? Is it safe to enjoy as a regular part of our diet?* Sorry. Pork should be avoided. The most common cause of adult-onset epilepsy is cysticercosis of the brain. Cysticercosis is caused by pork tape worm from the consumption of pork products. These products are also high in fats and salt.

- *Should phosphates in food be avoided?* Phosphates in food have been linked to kidney failure and premature death. Ironically, plants have more phosphates than animal products. However, the phosphates in plants are in a form that is not easily absorbed whereas the phosphates in animal products are readily absorbed. Phosphates are also deliberately added to soft drinks.

- *Should we have health concerns about consuming aluminum and Alzheimer's disease?*

  Dr. Chelliah counsels us that there may be a link between intake of aluminum and Alzheimer's disease. He advises us to avoid cheese because aluminum is intentionally added in large amounts to cheese commercially produced. Just one slice of individually wrapped processed cheese can contain up to 5 mg of aluminum.

- *Is it OK to eat fish?* Fish should be avoided because of toxins resulting from biological amplification—bigger fish eat smaller fish and accumulate toxins from the smaller fish which contain toxins from contaminated sea, lake, river, and other water sources where fish live.

- *What warnings, if any, can be given about consuming animal products?* All meats and animal products like eggs, milk, and cheese are associated with many systematic diseases including cancer, arthritis, heart disease, and stroke.

- *Is not chicken the safest meat to eat?* Regrettably, a virus is present in chicken meat that has been documented to cause breast and liver cancers. Endotoxins from bacteria in the chicken meat have been implicated in the causation of arthritis. Chicken also contains a large amount of fat and cholesterol. There was a fascinating study which showed that people who consumed chicken regularly had the highest amount of weight gain as compared to people who consumed other animal meats! Nabal contradicts this advice. He says: "Ah, hogwash! The founder of KFC died at 94 years of age. My grandpa drank warm beer, smoked his own rolled cigarettes, ate horribly smelly cheeses, and downed all the fried chicken that he could. When he was 78, he shouted at his critics: 'Don't count me out too soon!' He lived to be 100 and was trim at 100. So, don't tell me not to eat chicken!" Dr. Chelliah answers: *"The food consumed 100 years ago is totally different from similar food consumed today. For example, a serving of chicken 100 years ago contained 10 grams of fat. A similar serving of chicken today contains 27 grams of fat!! In addition, the commercially sold animal products available today have a lot more salt, fat, and industrial toxins."*

- *Can you alert us to any foods to avoid because they contain arsenic?* Commercially produced chicken feed contains arsenic. Chicken also is often contaminated with E. Coli bacteria which can cause severe illness when under-cooked chicken is consumed. Cooked chicken has heterocyclic amines which can cause cancer.

j Questions and responses pertaining to erectile dysfunction and prostate cancer:

- *What can you tell us about erectile dysfunction?* Let us assume that there are approximately 7.12 billion people on planet Earth and about 50 percent of all adults are male. We will assume that at least 2 billion, approximately, count for children. This would bring the calculation of men to around 2.6 billion. A reported estimate has it that about 100 million men in this world experience erectile dysfunction (ED). So, approximately 0.04% of men have erectile dysfunction. Amazingly, in the US, one estimate claims that there are 316 million people living in the US with about 124 million men of reproductive age. One reported estimate of US men with ED is 30 million. The stunning comparison follows. There are 30 million US males out of 124 million men of reproductive age that have erectile dysfunction. In the world, there are 100 million men with ED. The jolt follows: erectile dysfunction in men plagues 0.04% of them in the world versus 24% in the US alone. Conspicuously, the statistics on erectile dysfunction suggest that in the US, erectile dysfunction has reached epidemic proportions.

Let us analyze this further. Erection in men is brought about by an increased flow of blood into the penis and restricted outflow of blood causing engorgement of the penis. If there is poor blood flow into the penis, there is likely to be a problem with erection. Significantly, the blood vessel that supplies the penis is only half the size of the main coronary blood vessel that supplies the heart muscle. Subsequently, *the blood vessel that supplies the penis is likely to get clogged up long before the coronary blood vessel gets clogged up*. What has been shown in scientific studies published in reputable medical and nutrition journals is this. If a male has a problem with one blood vessel, he probably has a problem with the other. In other words, *a male with erectile dysfunction faces a great risk for developing a heart attack.*

Some experts have named erectile dysfunction "penile angina." *Others have referred to the penis as the "canary in the mine" since a penis with erectile dysfunction may warn of impending heart problems before a heart attack occurs.* Predictably, if a patient with erectile dysfunction undergoes a cardiac stress test, there is a greater than 80 percent chance that the stress test will unveil frightening results: namely, that the patient's coronary blood vessels are also already clogged. The point to be stressed is that erectile dysfunction and coronary heart disease are two different manifestations of the same underlying progressive disease—specifically, atherosclerosis or the clogging of the arteries. To err on the side of safety and sanity, cardiologist Dr. Chelliah's advice is that *men with ED should undergo a cardiac stress test.*

Erectile dysfunction inarguably diminishes the quality of life for both sexual partners and leads to depression in men. Mutually satisfying sexual activity characterizes a meaningful ingredient for success in sexual relationships. Sounding like Nabal, some men say, "Forget about it. Don't try to scare me. I'll eat whatever I like, and then I'll take Viagra for my erectile dysfunction." The problem with Nabal's approach is that Viagra, and similar drugs, only improve the situation temporarily and are associated with potentially serious negative side effects.

Although the ED situation looks just about hopeless, Dr. Chelliah maintains that *a wholefoods plant-based lifestyle offers the potential to reverse erectile dysfunction.* In one study, up to 37 percent of men with ED put on a vegan diet were able to reverse erectile dysfunction in 18 months. Temporarily improving erectile dysfunction can be achieved by eating watermelon, especially yellow watermelon, and by eating pistachio nuts. An interesting point related to women is that *a woman's sexual experience is also dramatically enhanced when she adopts a wholefoods plant-based diet.*

- *What can be said about the threat of prostate cancer?* In the Western world, prostate cancer is a major cause of illness and death among men. Globally, it is the second most common cancer. *Prostate rates in Asia can be as much as ten times lower.* Yet, Asians who move to and live in the US develop high rates of prostate cancer. In the US, nearly one in three men have small prostate cancers brewing, even in their 30s. In their 60s, nearly 2/3 have prostate cancer. On autopsy, most older men are found to have cancerous tumors in their prostates, cancerous tumors that went undetected. Amazingly, Asian men seem to have the same prevalence of latent prostate cancers on autopsy—cancers that never grew large enough to cause a clinical disease. In Japan, men tend to die with their tumors rather than from their tumors. *With Asians adopting the Western diet, their prostate conditions are worsening.*

In one study of foods that increase the growth and spread of prostate cancer, eggs and poultry respectively seem to be the worst. Eating one egg daily doubled the growth and spread of the prostate cancer. Eating one serving of chicken daily quadrupled the risk of cancer progression. The higher prevalence of cancers in Westernized Asians and Westerners may be due to multiple factors, including the consumption of large amounts of animal fats and meats, carcinogens in the Western diet, and an inadequate consumption of vegetables, fruits, soy products, and green tea. In a study on green tea, men drank six cups of green tea a day before undergoing prostate removal. The result was a 40 picomole cancer stopping concentration of green tea chemicals in their prostate tissues. Thus, the optimal amount of tea consumption appears to be six cups of green tea daily. *While green tea lessens the development of prostate cancer, green tea extracts do not work.*

Although a change in diet can benefit the health of the prostate gland, in spite of the calamitous effects of conventional clinical treatments of prostate cancer (including radical surgical resection, external radiation, and cryosurgery with erectile dysfunction being the most likely consequence), along with Nabal, most men do not want to change their diet and lifestyle. Almost a fourth of the men newly diagnosed with prostate cancer state that they would prefer to have their lives cut short rather than live on a diet that prohibits beef and pork. To be direct, *more men would rather be impotent than improve their diet.* Nabal, did you read the prior sentence?

After all is said and done, vegan diets seem to be the very healthiest. Nonetheless, going on a total vegan diet is easier said than done. Easier said than done though does not apply to the practical vegan diet since it can be achieved in gradual steps without grief or guilt. As Dr. Chelliah teaches us, the plant to animal (PM) ratio might be useful for cancer prevention in general as well as cancer prevention in the prostate gland. In the largest study ever performed on diet and bladder cancer, just a three percent increase in consuming animal protein was associated with a 15 percent higher risk of bladder cancer while a two percent increase in plant intake was associated with a 23 percent lower risk. Therefore, *even little changes in our diets can have significant health effects.* A practical vegan diet would constitute more than *"little changes"* with healthy benefits. And, for those who can savor a vegan diet, grand changes in health might follow.

Clinically malignant prostate cancer surfaces as being highest among Black-American males—some 30 times greater than among Japanese men and 120 times greater than in Chinese men in Shanghai. So, what can be eaten to improve a man's odds in avoiding prostate cancer? Generally, studies support drinking green tea, a vegan diet emphasizing cruciferous vegetables, and flaxseed consumption as possible preventions and even cures for prostate cancer. Being more specific, less than a single serving a day of cruciferous vegetables like broccoli, Brussel sprouts, cabbage, cauliflower, or kale may reduce the risk of cancer progressing by more than half—that is, the cancer coming back, spreading to bones, or death. Flaxseeds also fight prostate cancer. Garlic and onions appear to fight cancer as well. Cooked vegetables seem to work better than raw in the fight against cancer. Legumes (such as peas, beans, and lentils) protect against cancer too. Unfortunately for pomegranate juice lovers, a controlled trial of pomegranate juice revealed no beneficial effect on prostate cancer growth. *Flaxseeds in particular work against prostate cancer.* Researchers had men about to have their prostates removed eat three tablespoons of flaxseeds a day for several weeks before surgery along with a low-fat diet. The researchers were skeptical that they would observe beneficial differences with such a short period of dietary intervention. However, they found significantly lower cancer proliferation rates and significantly higher rates of apoptotic cell death—that is, cancer cell suicide. Note that *flaxseeds may have up to 800 times anti-cancer lignans than any other foods*. Flaxseeds can be served quickly and easily in cereals, smoothies, stews, soups, and other plant sources.

k Questions and responses pertaining to diabetes mellitus:

- *What can be said about diabetes and its treatment?* Diabetes is a dangerous and frightening disease that affects every organ in the body; it is the number one cause of adult-onset blindness, kidney failure, and surgical amputations. The US Centers for Disease Control and Prevention has estimated that by 2050, one out of every three of us in the US may have diabetes. Diabetes is caused by either not enough insulin being produced by the beta cells of the pancreas or by insulin resistance at the cellular level. Both *a low-fat diet and exercise* that burns the fat away can improve the interaction between insulin and the insulin receptors. In medical terms, this cellular phenomenon improves insulin sensitivity or reduces insulin resistance. When high fat food is consumed, a thin layer of fat covers the receptors and insulin is unable to find the receptors. The problem is not that there is a lack of insulin. The problem is that insulin is not able to find the receptors because of the fat covering the receptors.

  Two types of diabetes will be noted. Type 1 is diabetes mellitus. It is due to either the total absence or extremely low levels of insulin production by beta cells in the pancreas. Type2 diabetes mellitus is caused by a discrepancy between the amount of insulin needed and the insulin produced. In fact, in early stages of Type 2 diabetes, there is often an increased production of insulin because the actual insulin requirement is elevated way above normal due to obesity or insulin resistance. What is meant by normal in this context? The fasting blood sugar in a normal individual is less than 100mg/dL (deciliters). While a fasting blood sugar of 100-125 mg/dL qualifies for a diagnosis of prediabetes, a fasting blood sugar of >126 mg/dL qualifies for a diagnosis of diabetes.

Insulin is a chemical (hormone) produced by the "beta" cells scattered throughout the pancreas in small clumps ("islands") called *islets.* When the insulin producing beta cells are exposed to fat, *some of them* begin to die. An increase in blood fat levels is harmful to the pancreatic beta cells. There is a protein responsible for cell suicide—Programmed Cell Death Protein 5 (PDCD5). Some cells commit suicide through this protein. The protein appears to trigger cells, usually cancer or other abnormal cells to commit suicide: in short, they self-destruct. Regretfully, this protein can also destroy certain normal cells under certain conditions. For example, saturated fats (almost always from animals) can trigger the beta cells in the pancreas to commit suicide by increasing the levels of this protein. The unsaturated fat in olives, nuts, and avocados causes only a tiny bump in the level of this protein. In contrast, saturated fat from animals dramatically increases the level of this protein to promote beta cell death. Saturated fats and cholesterol (especially LDL) harm beta cells in the pancreas.

Diets rich in saturated fats cause obesity and insulin resistance. Also, increased levels of circulating non-esterified fatty acids cause beta cell deaths and may contribute to progressive beta cell loss and Type 2 diabetes. Infusion of fat into people's bloodstream and ingesting saturated fat can directly impair pancreatic beta cell function. Type 2 diabetes is characterized by defects in both insulin secretion and insulin-receptor interaction. Saturated fat appears to impair both. In summary, increased consumption of saturated fats has a powerful short term and long-term effect on insulin action, thereby contributing to the dysfunction and death of pancreatic beta cells in diabetes. Reducing the consumption of saturated fats may reverse this process. What causes diabetes is the consumption of too many calories in the form of saturated fats. To be direct, *fat causes diabetes, not sugar*. But of course, once you have diabetes, you will have to dramatically lower your intake of carbohydrates.

Of course, not everyone who consumes a high fat diet will develop diabetes. However, *eating a high fat diet is highly correlated with the onset of diabetes in many people.* Although all fats from animals increase the incidence of Type 2 diabetes, it appears that specific animal products may cause diabetes more than others. For example, six studies on the relationship between consuming fish and Type 2 diabetes have been published showing that fish eaters in the US tended to be at greater risk for diabetes. If Americans and Europeans are included in the analysis, fish eaters may have a 38 percent increased risk for the new onset of diabetes. In other words, there seems to be a five percent increase in the risk of diabetes for every serving of fish one eats per week. *Compared to red meat, a serving of red meat daily is associated with only a 19 percent increase in risk for developing diabetes. A serving of fish per day would amount to a 35 percent increased risk of developing diabetes.* The culprit appears to be the industrial toxins in fish.

So, how can fish be worse than red meat? Evidence commissioned by the US government revealed that blood sugars rose in diabetics given fish oil although it is not clear why fish oil increased the blood sugar levels in diabetics. A recent study found that the insulin producing beta cells in the pancreas do not appear to work as well in people who eat two or more servings of fish a week. It may be that the increased incidence of diabetes of people who consume fish may be correlated with the industrial (that is, human made) toxins or organic pollutants that fish carry: including, dioxins, PCBs, and some pesticides that are resistant to breaking down in the environment. Initially identified for their deleterious effects on reproductive function and their ability to cause cancer, there is now a growing body of evidence showing that exposure to these pollutants may lead to diabetes.

Although exposure to toxic pollutants in fish is strongly associated with an increased risk of diabetes, there is also a strong dose-dependent relationship between the levels of these pollutants circulating in the blood of the general population and diabetes. *People with the highest levels of pollutants in their blood stream had 38 times greater odds of developing diabetes*. While there is a correlation between obesity and diabetes, the correlation is only in those with high levels of industrial pollutants. Therefore, obesity was a risk factor for diabetes only if people had blood concentrations of these pollutants above a certain level. Important to note is that 90 percent of these pollutants detected in the general population result from animal products.

To keep diabetes from developing, what can we do? In a landmark study published in the *New England Journal of Medicine*, several thousand people were assigned randomly to receive a double dose of metformin, to participate in a lifestyle change in a diet and exercise regimen, or to receive a placebo. Compared to the placebo group, over the next four years, fewer people in the metformin drug group developed diabetes. Nevertheless, diet and exercise alone produced better results. *The diet and exercise group lessened the incidence of diabetes by 58 percent compared to just 31 percent with metformin.* Depending on the condition of the patient and the patient's circumstances, a physician might prescribe both metformin and the lifestyle change in diet and exercise. As Dr. Chelliah has told audiences repeatedly with respect to heart problems, *if a patient has clogged arteries, stents might be ordered first and then a diet high in plant foods accompanied by exercise.* The lifestyle change was significantly more effective than the drug and had fewer side effects. In closing, there is incontrovertible evidence that a low-fat vegan diet with vigorous daily exercise can prevent, reverse, and even cure a majority of Type 2 diabetes mellitus cases. Nabal disagrees with exercise with vehemence: "Malarky! Don't beat yourself up if you don't exercise. The creator of the treadmill died at age 54, and the inventor of gymnastics died at 57. So much for exercise, eh? Exercise just wears out the body. I think I'll take a nap on my couch right now!"

l Questions and responses pertaining to traditional medicine, modern pharmaceuticals, and spices—principally turmeric:

According to the World Health Organization, due in part to the high cost of modern medications, *80 percent of Earth's 7 billion inhabitants rely upon traditional medicine for their primary healthcare.* For centuries, medicines derived from plants have played an important role in caring for the health of people of all ages. *Spices comprise one important group of healing agents derived from plants.*

Known by different names around the world, turmeric is one such spice that has been used around the world for centuries. Nabal shouts: "Big deal. The stuff tastes terrible. Who can stomach it? The taste ruins me for hours. I can hardly get it past my lips. No thanks. Besides, anything that tastes that bad can't be good for me or anyone." We hear you, Nabal, but let us continue. Turmeric is the dried powdered root stalks of the turmeric plant. The turmeric plant belongs to the ginger family. From turmeric, the orangish-yellow pigment curcumin can be extracted. Curcumin gives turmeric its yellow *color.* In recent years, more than 5,000 articles have been published in the medical literature on curcumin. Many of these articles claim that curcumin can be useful in treating numerous conditions through various mechanisms. Significantly though, most of the 5,000 articles were merely in vitro (test-tube experiments) lab studies, not human subjects. Since the turn of the century, more than 50 clinical studies have been done that have tested curcumin against a variety of diseases.

Several findings are worthy of attention. Rheumatoid arthritis (RA) is an inflammatory disorder that causes progressive destruction of the cartilage and bone of joints. The long-term outlook of patients with RA is poor with as many as 80 percent of patients becoming disabled along with a reduction in years of life. *There are many drugs RA patients can take. Sadly, almost all of them cost the patient severe side effects*: such as, blood loss, bone loss, bone marrow suppression, and toxicity to the eyes and liver.

Curcumin versus phenylbutazone (a powerful anti-inflammatory drug) was demonstrated over 30 years ago in a double-blind crossover study to have the capacity to produce the desired result. The curcumin and the phenylbutazone groups showed significant improvement in morning stiffness, walking time, and joint swelling. However, the curcumin group enjoyed the complete absence of negative side effects. Because it damaged people's immune system and even caused death, phenylbutazone was removed from the market three years later. In another study, 45 patients with RA were randomly assigned to three groups: the curcumin group, the standard of care drug, or both. All three groups of RA patients improved. The *curcumin group though showed the highest percentage of improvement*, significantly higher than the improvement in the drug group and also safer in alleviating RA pain. In fact, there were more adverse reactions in the leading drug of choice group than in the combination of drug and curcumin group. Curcumin impressively reduced negative side effects of conventional medications used concomitantly. Moreover, *in contrast to nonsteroidal anti-inflammatory drugs, curcumin has no harmful gastrointestinal side effects and can even protect the lining of the stomach.*

Ingesting curcumin, the main ingredient of turmeric, is not the same as ingesting turmeric since turmeric probably contains several hundred other micro-nutrients that may collaborate with curcumin to enhance healthful benefits. An example of a collaborative beneficial effect of micronutrients working together is the adding of a pinch of black pepper to turmeric. *Just a pinch of black pepper increases the bioavailability of turmeric by a remarkable 2,000 percent.* By way of explanation, one teaspoon of turmeric when combined with a grain of black pepper has the same effect as consuming 20 teaspoons of turmeric.

An experimental study by researchers at the University of Florida and Penn State University that applied to spices and inflammation found that *cloves, ginger, rosemary, and turmeric significantly inhibited inflammation*. Black pepper alone, however, did not dampen the inflammatory response. Of course, there are drugs that can do the same thing—dampen the inflammatory response and lessen the suffering from inflammation related diseases like osteoarthritis, inflammatory bowel disease, psoriasis, and ankylosing spondylitis (a type of arthritis that may inflame at least the spine, pelvis, hips, and shoulders). More than 20 billion dollars a year in drug sales profit drug companies producing drugs that treat these diseases.

In contrast, curcumin in turmeric is a lot cheaper and safer way to treat the medical problems these diseases cause. The University of Florida and Penn State University study demonstrated that *cloves, ginger, rosemary, and turmeric significantly reduced the inflammatory response*. Consuming ginger for a week will decrease DNA damage. Including ginger or rosemary in our diet may lessen the damage by 25 percent. As for turmeric, this miracle spice can reduce DNA damage by 50 percent. In itself, curcumin is an antioxidant that boosts the activity of our bodily catalase antioxidant enzymes. When we take in approximately three quarters of a teaspoon of turmeric per day, the activity of the enzyme catalase in our bloodstream gets boosted 75 percent.

Furthermore, *turmeric works remarkably better than ginger and rosemary with a sizably smaller dose*. The researchers gauged one and a third teaspoons a day of ginger or rosemary to but a pinch of turmeric (about an eighth of a teaspoon) a day and observed comparable results. So, *to counteract inflammation, ginger and rosemary are good, but turmeric is much better*. Both raw and cooked turmeric are beneficial since one has some benefits which are not seen with the other. Researchers tested turmeric in both raw and cooked forms. They found that although raw turmeric had powerful anti-inflammatory effects, raw turmeric was not as effective in decreasing DNA damage as cooked turmeric. To repeat, the moral of this story is: *eat turmeric in raw and cooked forms*. Do you understand, Nabal?

Part 14: Summary and Conclusion/Take Away Ideas

In closing, we want to emphasize the following preferred motifs noted in this book for wholesome practical vegan diets. Although Nabal scoffs at these sensible dieting plans, we support them. For general purposes, Dr. Chelliah affirms that experience matters, so please choose experience—especially experience based on objective scientific measures with wisdom and as bias free as possible. For hearty health purposes, while a pure vegan diet may have more advantages than a practical vegan diet, in the US and Canada, a small number of citizens, perhaps 5 percent, may be able to embrace a lifestyle that renounces all meat and meat products. Ergo, a *practical* vegan diet at a high ratio can be attained by just about anyone and can still be highly beneficial. For a test or study that runs several months to a year, some omnivores may be able to tolerate and endure a vegan diet. Once the study ends, so dies the vegan diet. With practical veganism, the practical vegan beat can go on for life. A high ratio of plant to meat, perhaps a PM ratio of 70 to 30 or better 90 to 10, should be reached and maintained on a practical vegan diet. A low ratio of carbs to fiber should be a daily practical vegan goal: maybe a carb to fiber ratio of 5 to 1—not 25 or more to 1. When medical conditions demand it, a practical vegan or vegan diet should not substitute for surgical or pharmaceutical interventions first. A "practical" vegan or "impractical" total vegan diet should follow the medical treatment.

For purposes of vitality and elan, *we should eat to extend our healthspan as well as our lifespan.* When conditions prevail, we should choose to eat natural plants rather than the juice or oil from natural sources: for example, cherries rather than cherry juice, carrots rather than carrot juice, and almonds rather than almond oil. For heart health, when conditions permit, we should select beans and lentils to ward off hunger and lose weight. It is wise *not* to assume that all vegetables are healthiest when raw. As a reminder, mushrooms and tomatoes are healthiest when cooked. While Dr. Chelliah encourages everyone to live a total vegan lifestyle, he knows the practice of total veganism constitutes a barrier for most people and oftentimes results in failure thanks to, among other forces, a low motivation to satisfy the requirements of a pure vegan lifestyle. He does not want to scare people away from pure veganism with a tyranny of *shoulds* and *should-nots*. To be pragmatic and reach a myriad of dieters, Dr. Chelliah offers the world *practical* veganism. While practical vegans may never aspire to total veganism, practical veganism can move anyone toward complete veganism. We have made a supreme effort to avoid politicizing and commercializing dietary advice on practical veganism and veganism. In brief, despite Nabal's naysaying nonsense, we offer a sane and progressive alternative to total veganism. The practical vegan diet we endorse is one with plant to meat ratios gradually advancing from PM ratios of 10:90 to PM ratios of 90:10—or even more if one's constitution, willpower, and discipline allow. Still, the practical vegan diet must be tasty. So, we choose to end our book with this refrain: *Good food must first get past our lips to do us any good!*

NOTE

1. The following selection of scientific articles address diverse aspects of veganism:
   A. Benatar, J. R., & Stewart, R.A.H. (2018). Cardiometabolic risk factors in vegans: A meta-analysis of observational studies. *PLOS ONE*, 13 (12): e0209086.
   B. Craig, W. J. (2009). Health effects of vegan diets. *The American Journal of Clinical Nutrition,* 89 (5): 1627S–1633S.
   C. Craig, W.J., & Mangels, A.R. (2009). Position of the American Dietetic Association: Vegetarian diets. *Journal of the American Dietetic Association*, 109 (7): 1266–1282.
   D. Di Genova, T., & Guyda, H. (2007). Infants and children consuming atypical diets: Vegetarianism and macrobiotics. *Paediatrics & Child Health*, 12 (3): 185–188.
   E. Gan, Zuo Hua et al. (2021). Association between plant-based dietary patterns and risk of cardiovascular disease: A systematic review and meta-analysis of prospective cohort studies. *Nutrients,* 13 (11): 3952.
   F. Greger, M. (2015). *How not to die: Discover the foods scientifically proven to prevent and reverse disease.* New York City: Flatiron Books.
   G. Huang, R. et al. (2015). Vegetarian diets and weight reduction: A meta-analysis of randomized controlled trials. *Journal of General Internal Medicine*, 31 (1): 109–116.
   H. Iguacel, I. et al. (January 2019). "Veganism, vegetarianism, bone mineral density, and fracture risk: a systematic review and meta-analysis". Nutrition Reviews. 77 (1): 1–18.
   I. Kaiser, J. et al. (2021). "A systematic review of the association between vegan diets and risk of cardiovascular disease. *The Journal of Nutrition*, 151 (6): 1539–1552.
   J. Medawar, E. et al. (2019). The effects of plant-based diets on the body and the brain: A systematic review. *Translational Psychiatry*, 9 (1): 226.
   K. Ornish, D. (1990). *Dr. Dean Ornish's program for reversing heart disease*. New York: Random House.
   L. Ornish, D. (2008). *The Spectrum*. New York: Ballantine Books.
   M. Quek, J. et al. (2021). The association of plant-based diet with cardiovascular disease and mortality: A meta-analysis and systematic review of prospect cohort studies. *Frontiers in Cardiovascular Medicine*, 8: 756810.
   N. Rees, K. et al. (2021). Vegan dietary pattern for the primary and secondary prevention of cardiovascular diseases. *The Cochrane Database of Systematic Reviews*, 2021 (2): CD013501.
   O. Rizzo, G. et al. (2016). "Vitamin B12 among vegetarians: Status, assessment and supplementation. *Nutrients (Review)*, 8 (12): 767.
   P. Sanders, T. A. (1999). The nutritional adequacy of plant-based diets. *The Proceedings of the Nutrition Society,* 58 (2): 265–9.
   Q. Smith, A. M. (2006). "Veganism and osteoporosis: A review of the current literature. *International Journal of Nursing Practice*, 12 (5): 302–306.
   R. Stone, G. (Ed.) (2011). *Forks over knives: The plant-based way to health*. New York City: Workman Publishing.

S. Turner-McGrievy, G., & Harris, M. (2014). Key elements of plant-based diets associated with reduced risk of metabolic syndrome. *Current Diabetes Reports*, 14 (9): 524.

T. Tuso, P. J. et al. (2013). Nutritional update for physicians: Plant-based diets. *The Permanente Journal*, 17 (2): 61–66.

REFERENCES

Burr, M.L. et al. (1989). Effects of changes in fat, fish, and fiber intakes on death and myocardial reinfarction: Diet and reinfarction trial. *Lancet*, 2 (8666): 757-761. [Referred to in this book as the DART-I Trial].

Burr, M.L. et al. (2003). Lack of benefit of dietary advice to men with angina: Results of a controlled trial. *European Journal of Clinical Nutrition*, 57 (2): 193-200. [Referred to in this book as the DART-II Trial].

Burrell, B. (1996). *Merriam-Webster's pocket guide to business and everyday math*. Springfield, MA: Merriam-Webster, Inc.

Chelliah, N. N. (2022). Brochure on heart health. Grand Forks, ND: Heart Institute of North Dakota; and, for all of Dr. Chelliah's *Health Organization* brochures contact Dr. Chelliah at drchelliah@heartnd.com.

Christensen, C.M., Grossman, J.H., & Hwang, J. (2009). *The innovator's prescription: A disruptive solution for health care*. Toronto: McGraw Hill.

Conchie, B.E. (1975), *Business and consumer mathematics*. Don Mills, ON: Addison-Wesley (Canada) LTD.

Craig, W. J. (2009). Health effects of vegan diets. *The American Journal of Clinical Nutrition*, May, 89 (5): 1627S–1633S.

DeVito, J. (2021). Interpersonal time: A neglected dimension. *ETC: A Review of General Semantics*, 78 (1 & 2).

Eliade, M. (1978). *A history of religious ideas: From stone age to the Eleusinian mysteries*. Chicago: The University of Chicago Press.

Fiordo, R. A. (1977). *Charles Morris and the criticism of discourse*. Bloomington: Indiana University Press and Lisse: The Peter de Ridder Press.

Fiordo, R. (2018). Judgment day for general semantics: A judgmental orientation, *ETC: A Review of General Semantics*, 75 (1 & 2): 104-161.

Hargreaves, S.M., Raposo,A., Saraiva, A., & Zandonadi, R.P. (2021). Vegetarian diet: An overview through the perspective of quality-of-life domains, *International Journal of Environmental Research and Public Health,* April 12, 18(8): 4067.

Horney, K. (1950). *Neurosis and human growth*. New York: Norton.

"How To Do Peripheral Heart Action (PHA) | Training System" at https://youtu.be/nyv9iSh16Z4 (May 10, 2020) on *youtube.com*.

Kahneman, D. (2011). *Thinking fast, thinking slow*. New York: Macmillan.

Katcher, H.I., Ferdowsian, H.R., Hoover, V.J., Cohen, J.L., & Barnard, N.D. (2010). A worksite vegan nutrition program is well-accepted and improves health-related quality of life and work productivity. *Annals of Nutrition and Metabolism*, 56(4):245-52.

Kornfield, J. (Ed.) (1993). *Teachings of the Buddha*. London: Shambhala.

Kruger, J., & Dunning, D. (1999). Unskilled and unaware of it: How difficulties in recognizing one's own incompetence lead to inflated self-assessments. *Journal of Personality and Social Psychology*, 77(6): 1121-1134.

Piras, A., Persiani, M., Damiani, N., Perazzolo, M., & Raffi, M. (2015). Peripheral heart action (PHA) training as a valid substitute to high intensity interval training to improve resting cardiovascular changes and autonomic adaptation. *European Journal of Applied Physiology*, 115 (4): 763–773.

Morris, C. (1964). *Signification and significance: A study in the relations of signs and values.* Cambridge, MA: M.I.T. Press.

Mueller, L. (2021). Minimize inflammaging and maximize your healthspan with turmeric. *good4u Health Hotline*. July, Vol. 48.

Safrit, M. J., & Wood, T. M. (1995). *Introduction to measurement in physical education and exercise science, 3rd ed.* London: Mosby.

"SERGIO OLIVA'S FIRST BODYBUILDING ROUTINE-PHA SYSTEM! BLUEPRINT TO THE MYTH!!" at https://youtu.be/ogABJLruQu8 (April 15, 2019) on *youtube.com.*

Thomas, A. (1993). *The vegetarian epicure: Book two*. New York: Alfred A. Knopf.

Thomas, J. R., Nelson, J. K., & Silverman, S. J. (2005). *Research methods in physical activity, 5th ed.* Champaign, IL: Human Kinetics.

Violato, C. (2019). *Assessing competence in medicine and other health professions*. London: CRC Press.

Weaver, R. M. (1948). *Ideas have consequences*. Chicago: University of Chicago Press.

Wikipedia.com (2021). Sub verbo: cuisine, *omnivore, vegetarian, vegan, carnivore, herbivore, flexitarian, Big Texas Roadhouse, Heart Attack Grill, PHA training, heterocyclic amines.*

www.nutritionalfacts.org

ACRONYMS (ABBREVIATIONS)

ARMD	Age-Related Macular Degeneration
CBDE	Competency Based Dietary Education
CBME	Competency Based Medical Education
CDC	Centers for Disease Control and Prevention
DART	Diet and Reinfarction Trial
DHA	Docosahexaenoic Acid
DII	Dietary Inflammatory Index
DNA	Deoxyribonucleic Acid
ED	Erectile Dysfunction
EDA	Entrustable Dietary Activities
EPA	Entrustable Professional Activities
IGF-1	Insulin Like Growth Factor 1
HDL	High-Density Lipoprotein (so-called "good cholesterol")
LDL	Low-Density Lipoprotein (involved in atherosclerosis)
MP	Meat-Plant, Meat to Plant, or Meat: Plant
PCB	Polychlorinated Biphenyl (an environmental toxin)
PDCD5	Programmed Cell Death Protein 5
PHA	Peripheral Heart Action
PM	Plant-Meat, Plant to Meat, or Plant: Meat
PMS	Premenstrual Syndrome
PV	Practical Veganism
RA	Rheumatoid Arthritis
SAD	Standard American Diet
WHO	World Health Organization

INDEX

Made in United States
Orlando, FL
04 November 2026

90202677R00070